# PEEP
# LIGHT

# PEEP LIGHT

Stories of a Mississippi River Boat Captain

## Lee Hendrix

University Press of Mississippi / Jackson

The University Press of Mississippi is the scholarly publishing agency of
the Mississippi Institutions of Higher Learning: Alcorn State University,
Delta State University, Jackson State University, Mississippi State University,
Mississippi University for Women, Mississippi Valley State University,
University of Mississippi, and University of Southern Mississippi.

www.upress.state.ms.us

The University Press of Mississippi is a member
of the Association of University Presses.

Library of Congress Cataloging-in-Publication Data

Names: Hendrix, Lee (Leo), author.
Title: Peep light : stories of a Mississippi River boat captain / Lee
Hendrix.
Description: Jackson : University Press of Mississippi, 2024. | Includes
bibliographical references.
Identifiers: LCCN 2023035288 (print) | LCCN 2023035289 (ebook) | ISBN
9781496848185 (hardback) | ISBN 9781496850362 (trade paperback) | ISBN
9781496848178 (epub) | ISBN 9781496848161 (epub) | ISBN 9781496848154
(pdf) | ISBN 9781496848147 (pdf)
Subjects: LCSH: Hendrix, Lee (Leo) | Ship captains—Mississippi
River—Biography. | Steamboats—Mississippi River. | Mississippi
River—History.
Classification: LCC VK140.H46 A3 2024 (print) | LCC VK140.H46 (ebook) |
DDC 386/.35092 [B]—dc23/eng/20231031
LC record available at https://lccn.loc.gov/2023035288
LC ebook record available at https://lccn.loc.gov/2023035289

British Library Cataloging-in-Publication Data available

To Captain Clarke "Doc" Hawley, the greatest riverlorian of them all.
If I could only tell stories the way he did . . .

And to Captain Adrian Hargrove. I hope you've found the water
that's deep and slack. Just save some for me. Without your guidance,
I couldn't have done any of this.

# CONTENTS

# Contents

# ACKNOWLEDGMENTS

I would first like to thank my wife, Dianne Coombs, for her help over the years with computers and creating files. Without her looking over my shoulder and running me off the keyboard when the Mac world got too complex and frustrating for me, I would never have been able to put these thoughts into any sort of comprehensible form. Likewise, my gratitude goes to the University Press of Mississippi for finding artistic value in the stories of a river rat and particularly to Lisa McMurtray and Craig Gill for guiding me through a process that I was completely unfamiliar with. Special kudos to copy editor Laura J. Vollmer for being so accessible. St. Louis folks are just that way.

Of course, if Don, Bob, and Les had never put me on my first towboat and, miraculously, seen fit to make me a pilot, there wouldn't ever have been too much for me to write about. They led a loyal and understanding company that was easy to work for. Watching and steering for Jimmy Bryson, Leroy Drury, Buddy Compton, Norris Frederick, Sam Hanks, Ray Puckett, and many others were invaluable experiences for my development and gave them a bit of a break too, I hope—when I wasn't scaring them to death. I also listened to their stories, many of which were quite memorable, and a few others, well, quite unprintable.

The contributions of Butch Harrington cannot be overstated. He was always there to challenge me into getting my work out, and he read just about anything I ever sent him. John Dugger gave me tales about the Army Corps of Engineers, keeping me so entertained that I vowed I would someday experience it myself. Cary Lewis and I spent many hours together in the pilothouse of the *M/V Mississippi*, talking about the mom-and-pop companies we had both worked for. He was always prompt in reading my stories and suggesting other books for me to read, as were ex–river rats Bill Rathbun and M. R. Rothschild. No matter how few years one works on towboats, or how many years ago it has been, those memories never seem to fade. Tom Dunn, Dick Karnath, Carl Henry, and Dr. Arnie Jacobson (in addition to working on my teeth) always provided feedback and kept the storyline moving.

There was Gary Frommelt, who introduced me to the passenger boat industry and to Doc Hawley. John Hoover, Charles Brown, and Sara Hodge at the Mercantile Library and the late David Lobbig at the Missouri History Museum were there to help me with research and other special projects. John David Holmes and I spent long hours "walking in Memphis" and talking about writing. I wish him the best in his current work.

Reggie, Molly, and Ben at *Big River Magazine* were the first ones to value my work enough to put it in print and help me with other stories about the river. Steven Marking provided me with hundreds of images and kept me enthused about the pursuit of bringing the river to life for boat passengers. Frank Rivera influenced me to give the riverlorian gig a shot, as did ex-riverlorians Clara Isenhour and "Toots" Maloy. Sue Kruger and the Lake City Storyteller's Collaborative always politely listened to everything I presented. "Minnesota nice" really does exist in this cruel world.

I can never forget the educators I worked with that provided me with a different vision of life regarding the river and the natural world. Hank Schafermeyer convinced me to briefly change careers in my mid-thirties and to see that I could make a difference in the lives of others. Tom Ball, Vic Nelson, and Byron Clemens (a distant relative of Sam) were among the many teachers that I shared a different type of river experience with.

I recognize my mom for worrying about me so much while I was on the river, my dad for taking me to the river in the first place on October 9, 1972, and my sister, Sally, for coming out to the boats to visit and bring me T-shirts. And I recognize Mitch for succeeding in life, even though his dad was gone for weeks on end.

I'd like to give artistic thanks to Hank Williams (Sr., Jr., and the III), Patsy Cline, the Kings (B.B., Albert, and Freddie), Social Distortion, Rancid, the Stone Temple Pilots, Cody Jinks, Soundgarden, Sublime, Elmore James, Lucero, Tyrone Davis, George Jones, The Temptations, and all the other various artists that helped keep me awake and revved up in the pilothouse all these years. And I will never forget Dave Williamson and I jamming on "Radar Love" and "Dark Side of the Moon" in the pilothouse of the *American Queen*. I always wondered if the passengers and Captain Keeton had any idea what we were doing up there at 2 a.m.

In the category of too many to thank are all the folks I've worked with on towboats, casinos, steamboats, and in the corps for fifty years. My homies on the *M/V Mississippi* and the *American Queen* stand out particularly. I will serve with them every chance I get.

# RIVERBOAT GLOSSARY

**barge coaming:** Dry-cargo barges have this superstructure a few feet off the deck, just at the perfect height to snag your life jacket and poke you not too gently in the side.

**batture:** This is the alluvial land between the riverbank and the levee.

**break coupling:** On a twelve- or fifteen-barge tow, the break coupling is two tiers of barge length positioned out from the towboat. Simply put, this is where the tow is broken in two halves for a double lock.

**bumper:** Made to soften the impact between barges and a lock wall, a bumper is usually woven with old pieces of line.

**capstan:** This is a windlass or a mechanical aid (usually electric) for tightening lines on the foredeck of the towboat. It is commonly used for "facing up" prior to putting face wires on. If you put a glass top on it, it becomes an engaging cocktail table.

**cavel:** The spelling of this one is tricky. In proper nautical parlance, it is spelled "kevel." However, we river rats have our own dialect. Some yachtsmen and other marina folks call it a "cleat," but those are usually a much smaller object. Anyway, lines are made fast on these anvil-shaped pieces of heavy metal that are welded onto the deck of boats and barges.

**cheater bar/cheater pipe:** Used for leverage to tighten a wire, they are usually about three- to four-foot-long hollow pipes.

**deadman:** I am smart enough to not give the story away. You have to read it.

**deckhand:** As in "poor, old deckhand," they are the backbone of the industry. They are out locking and putting barges together in minus-ten- or over-one-hundred-degree weather. They are also responsible for the cleanliness of the boat. Good deckhands are a treasure and should be appreciated by everyone.

**dikes (wing dams):** These are formations, usually made of rock, to direct the flow of the river where the Corps of Engineers wants it to go. Traditionally, they were set perpendicularly in a straight line from the bank, but now come in various shapes (such as horseshoes).

**double locking:** Since many tows are longer than some locks, the tow must be locked in two (or very infrequently more) segments. This is a standard practice on the Upper Mississippi River where most locks are six hundred feet long while many tows approach one thousand feet in length.

**double tripping:** This is the hateful practice of breaking a tow in two because the towboat is incapable of shoving up through a particularly tight or swift place in the river. One part of the tow is tied off while the other part is taken through, then the towboat goes back to get the other part. This method has gotten less and less commonplace over the years and is now rarely employed because boats are better and because deckhands can quit their jobs and get home easier than they used to be able to.

**face wires:** These are stout riggings that hold the towboat to the stern of the tow. They are usually tightened up with an electric winch mounted on the forward deck. When I was a deckhand, face wires were made of steel. Today, they are made of a lighter material and are still just as strong. "Facing up" the boat means attaching the boat to the tow.

**flanking:** This is a maneuver during which a pilot reverses engines to bring a tow down to the speed of the current and floats around a heinous river bend. It is used as a safer alternative to trying to steer a tow around the bend.

**fleet:** Barges tied off to the bank or on anchors awaiting a towboat to take them up or down the river.

**fore and afts:** These are usually wires, but could be lines, that secure couplings of barges together. They are attached longitudinally on the ends of barges.

**handy line:** This is a line with a thin diameter, usually not much bigger than a clothesline or a canoe painter, used to pull a larger line to the shore or up on a lock wall.

**jockey wires:** Don't ask me why they're called that. That's just their name. They do hold barges together side by side and keep them from sliding back and forth.

**lines:** On my first day, I was told, "Cowboys use ropes, rivermen use lines." Surprisingly, there are still a few veteran rivermen who refer to lines as "ropes." Previously made of manila, today most are made of a synthetic blend that can be very dangerous to those standing close to them.

**lock on the head:** This is a riverman's vernacular for "Boys, it's time to get ready to make a lock."

**mate:** The mate is second in command of the vessel. When I first started, the mate led the deck crew in all activities and tasks. This position has evolved, legally, into the pilot role. In practice, many towboats still adhere to the traditional mate role.

**monkey fist:** This small handy line is wrapped around a solid object, such as a ball or another weight. It then becomes like a ball of twine, only heavy enough to hurl it from a lock wall or a barge to a waiting deckhand who usually ties it into a larger line. It is a method of transporting a line when the line itself is too awkward or heavy.

**mule training:** This is an antiquated method of getting barges through ice. The barges are put in a long single-file line behind the towboat and pulled along with "ice lines" instead of hard rigging. Usually, one load was faced up to the towboat to break through the ice. This strategy, as far as I can tell, is no longer employed due to danger to the deck crew and its impractical nature. I did, however, mule train pontoon barges while working on the Dredge Goetz just a couple of years ago (not in ice). It was a reminiscence but not a happy one.

**revetment:** This stabilizes the riverbank to prevent "cutoffs" and "blowouts." On most rivers, rock is employed. On the Lower Mississippi, the Corps of Engineers sinks concrete "mats," which are fabricated at "mat yards," at the designated bank using large gantry cranes.

**riding shotgun:** When a "first cut" is pulled from a lock by an electric winch, two people are tasked with the dangerous job of stopping it, armed only with lock lines. The mate is usually on the head of this steel phalanx, while a trusted deckhand is on the stern "riding shotgun."

**rigging:** This refers to hard steel that holds barges stiffly together. Portable rigging consists of a ratchet, wire, and sling, all feeling as if Stone Cold Steve Austin is pulling on the other end as you "wrastle" with it. More and more commonly, "stationary rigging" is welded into the barge decks, which is not nearly so hernia producing as the original steel jewelry.

**Rose Point:** This is a brand of an electronic navigation system employed on most Western Rivers towboats. It has revolutionized piloting, so Neckbone (from my story), or any other rookie pilot, would find it more difficult to get lost. It even predicts ETAs and closest points of approach (CPAs) with other river traffic. Most young pilots today have never navigated without it.

**souging:** This is a term only used on river boats, as far as I can tell. No one else has ever heard of it. It refers to cleaning the boat, inside and out, even in the engine room, with soap, water, and rags. "Git them boys to sougin'" the captain would say if he saw us enjoying the scenery too much.

**splice:** This refers to weaving strands of line to form an eye or put two pieces of line together.

**timberhead:** In seafaring lingo, these columns of heavy metal are known as "bollards." They are mushroom shaped and come in pairs, welded into decks of boats and barges. They are utilized to secure lines.

**toothpicks:** These are solid steel bars put into ratchets to tighten them. They look like a police nightstick and weigh about ten pounds. Not recommended for dental hygiene.

**towboat:** A towboat is a boat suitable for pushing barges ahead of it. Many ask me, "How is it living on a barge?" And I smile. The crew does not live on the barges. Living quarters are on the towboat. Barges are normally mere hitchhikers, though some towboats do have "dedicated" barges (usually liquid cargo).

**towing line:** Or it could be a wire. The key element that differentiates it from a "backing wire" or "backing line" is how it leads. The towing wire leads back from the boat or the string of barges directly attached to the towboat so you can "tow" the other strings along. Logically, a "backing line" does the opposite.

**tow knee:** This is a large column of metal welded onto the front deck of a towboat. Hard rubber or some other contact material is attached to the tow knee so the towboat can "face up" to barges and push them without sliding back and forth. Some tow knees have caches of rigging inside them. Tow knees are tall enough to face up to empty barges as well as loads.

**watch:** This is the shift that the crew works. Traditionally, there are two watches of six hours, though some are trying different schemes to relieve fatigue.

**wheels:** These are what recreational boaters call "propellers." The usage of these words on the river is usually how you can tell a work boater from a party boater.

**wing wires:** These are more face wires for holding the boat steady, usually strung to the outermost-face barges. Because of their length, they are harder to deal with and can be treacherous on slippery decks or when deployed up to empty barges. It used to be a bone of contention as to which watch would "get to" put the wing wires out.

**yawl:** This is a small boat used to carry a line to shore, make a crew change, or pick up supplies. It is normally the rescue boat as well. Today, most have motors, but previously, yawls were rowed. We often had jon boats that were used as yawls.

Finally, the question everyone asks: What's the difference between a **tugboat** and a **towboat**?

It's kind of complex. "Towboat" is a pretty specific term implying a workboat with tow knees that pushes barges in front of it. On the Western Rivers, it also implies a live-aboard crew. Tugboats, on most of the Western Rivers, imply smaller towboats that assist larger towboats with fleeting, arranging barges in the tow, or with locks. In addition, below Baton Rouge, tugboats are a completely different breed. They do not have tow knees and therefore cannot push barges in the traditional sense. They have rounded bows and hulls, making them suitable for open water. They are used to assist large ships into a berth, or they can pull barges behind them in open water. Near the Gulf of Mexico, boats with tow knees are sometimes referred to as "push boats." I hope I cleared that one up for the reader because I still get confused.

And, of course, **bow** is the front, and **stern** is the back. **Starboard** is right; **port** is left.

# PEEP LIGHT

# PROLOGUE

On our first day of class in 1965, my high school English teacher proudly informed us that she had once given Tennessee Williams a failing grade. Suitably intimidated, I have waited over a half century to attempt publishing a book. I finally feel safe to give it a try because I think she is probably deceased by now.

It has often been said that if a riverman ever tells the complete truth, he loses one of his ears. Last time I checked, I still had both of mine. What follows is what took place, to the best of my recollection. My mission is to bring you to the crossroads of history and legend. The rest is up to you, my friend. Some names have been changed for the benefit of those that deserved it—and for some that probably didn't.

## CHAPTER 1

# Between the Sticks

He was accustomed to it, being his own best friend, alone in the darkness of a pilothouse. Almost all towboats, when he had started a half century before, had long steering and flanking rudder levers protruding from the console. When you assumed your watch, you were "between the sticks." Now, some of them have joysticks, Z-drive units, wheels, or other assorted gadgets by which to steer a boat. He had even worked on a boat once that had a bicycle chain across the console, displaying the rudder angle. A traditionalist, he still preferred to say he was between the sticks.

Some things change; others don't. The time 0200 is still 0200. Anybody with any sense, or choice, was sound asleep. The folks at the office sure were, and they were not worried at all about him—they never were unless he caused them some problems. The old-time steamboat pilots used to call a night like this "dark as the inside of a cow"—not that the inside of a cow would be any darker than the inside of, say, a giraffe.

But the old-timers knew a lot, and if you were smart, you took notes. If you weren't smart, you learned like most people do—the hard way. He hadn't always been so smart.

Last month, he heard a couple of deckhands refer to him as "that old man." He chuckled to himself in the blackness of the night, then muttered, "Yep, I am. But with a young man's heart. Still in search of something that I may never find." You have to talk to yourself a bunch more now than you did back then because nobody wants to come up to the pilothouse and talk at 2 a.m. like they did years ago. Some of them have said they want to become pilots, but most would rather watch TV or play iPhone games than sit up there in the dark with a seventy-year-old man. Staying up while off watch to talk to the pilot or captain was how he had learned. Some things change; others don't.

There was no moon up around Morganza that night, but soon that wouldn't matter. In less than an hour, he'd begin to see the lights around St. Francisville, later the "Exxon glow" in the sky above the tree line that foreshadowed arrival into the "combat zone" at Devil's Swamp, then on down to Baton Rouge. He'd have to switch the VHF radio to Channel 67, ending the relative gentleness of the night. In the meantime, he'd keep following the peep light. The little blue light on top of the jackstaff pole, out there several hundred feet, would be his beacon, as it had been on so many nights.

He hoped the captain, his partner, was sleeping well so that he would be relieved on time by a fresh face. Then he could go down to his room and hope for just a few good hours of sleep—four and a half if he was lucky, four hours and fifteen minutes if he was not. Sleep is the most valuable commodity on a towboat, even more than food, and is measured in razor-thin margins. He would try to make it down through the upper bridge for his partner, but it would be close.

It was time to quit the caffeine, or he'd never get to sleep. In the old days, he might have lit up a Lucky Strike. In the daylight back then, the pilothouse walls shone yellow from nicotine. Not anymore; no indoor smoking is allowed. He could have run outside and smoked, but he would have had to hurry back in before the "dead-man alarm" woke everybody up. He especially didn't want to wake the captain up. Hell, his knees and hips were so stiff, he might fall trying to run back inside. It wasn't worth the risk. Some things change; others don't.

The electronic chart off to his side was a useful and friendly gadget. When he had first heard of it, he thought it would just make a handy jacket rack because a good pilot wouldn't need it. A real pilot had to have the river inside his head. Now, if the Rose Point chart went down, he wanted to pull over and stop. When he broke in, so many years back, all he had was a large radar set that pulsated eerily as the image swept across the screen. Occasionally, it broke down, but the office had expected him to keep going anyway.

Well, the office did care enough, now, to get him Sirius radio, and he put on *Willie's Roadhouse*. Even in the old days, no matter where you were, you could always pick up a country station. Some things change; others don't.

Often, on nights like this, it occurred to him to retire. He would miss parts of it. The solitude, strangely enough. It was easier to be an observer of the rest of the world and offer commentary than to be a participant. And the sunrises and sunsets, though the back watch doesn't see many of them. Of course, the money hose was much wider than it was in 1972 when he had made twenty-five dollars a day. It's tough to give up the money hose; you may never get to hold it again. But it was more fun back then, rowdy and raucous, and every towboat had a cook and two engineers. That thought brought another smile to his lips because most things are more fun when you're twenty-five, though the isolation had felt worse then.

There was plenty of high water to think about. His first spring, 1973, had challenged the apocalypse of 1927 for flood records. The captain had stuck him, a green deckhand, on top of the pilothouse when they had crawled up through the Eads Bridge in St. Louis to see if it would clear. He had felt vital to the mission as he talked to some stranger, long haired and young, like he was then, who was walking over the bridge's rusty steel arches to Illinois. They had cleared the low steel by a few inches. Later that same day, upon reflection, he had thought about what would have happened to him if they hadn't cleared. There was 1993 when they had to move the gambling casino from landing to landing, and no towboats even went through the Eads Bridge because it was so perilous. And of course, 2011, when all the records were shattered and the previously unimaginable had happened. He had had a hand in blowing the levee up at Birds Point to save the others in the system.

Yes, and there were always women. Late-night country music is sure to take you there. Back then, at 2 a.m., he worried she was cheating on him, planning to leave him for the electrician down the street, who was always home. Now, he's married to someone else and worried about her blood tests and if she's safe and comfortable at home with the Lab at her side. Worrying about the electrician was somehow easier on the mind. Willie came on the radio: "Mamas, don't let your babies grow up to be cowboys."

In the next stanza, he sang along, but changed a few words: "Mamas, don't let your babies grow up to be river pilots."

Some things change; others don't.

His mind wandered back to his childhood. He was again burning leaves from the gutter with Papa and listening to Nana's stories. His life story reset and turned to seeing the river for the first time, to being taught to fear it. Strangely, he had found himself wanting to be part of it. Suddenly, it was 1972 all over again, the first of many towboats. He had placed the jackstaff out on the head of the tow for the pilot to check his swing, with the peep light on the top of it for nighttime. The peep light had led that pilot and him through thousands of dark nights, like this one. He had no choice but to keep following it.

# CHAPTER 2

# One Good Deckhand Can Pull a Dead Man from the River

A St. Louis native from the nineteenth century, T. S. Eliot, saw the Mississippi River and declared it "a strong, brown god, sullen, untamed, and intractable." That personification remains an accurate reflection of how most in my birthplace view the Mississippi. The river there is feared like an unforgiving and vengeful deity from the Old Testament. As a child, I was carefully ushered to the Mississippi by my parents and ordered to stay out of its murky, churning waters. In case that warning wasn't sufficient, I was also taught to consider the river as a child abductor that would steal me away, murder me, and trap me for eternity in a muddy tomb.

Almost no one chooses to immerse themselves in the Mississippi in St. Louis, unless they are bent on attempting suicide. Notwithstanding, I found myself drawn to it. I respected the Mississippi's power, of course, but I also dreamed about it as an escape path. The river had everything that I lacked. It was majestic, powerful, beautiful, and free. Like Tom Wingfield in *The Glass Menagerie*, I wanted to escape from St. Louis and seek adventure. Or so I thought.

Luckily, St. Louis was the epicenter for the river industry in those days. As I looked in the yellow pages, I saw some entities that were known as "towing companies." Damn, I didn't want to work on a tow truck even if it was on water! How did that play out? Some called themselves "barge lines." That seemed much more in line with logic. On applications, I didn't even know what to label my own ambition, writing that I wanted to be a "barge laborer," which at least sounded appropriate. The appellation of "deckhand" was not even in my vocabulary.

I went on a couple of interviews at plush offices in Clayton, and finally, a representative from a "towing company" called me. I guardedly accepted the job since it was the only offer I had gotten. I asked him where I needed to go, and he said, "Be at the Missouri Portland Cement Dock on Riverview Boulevard." I asked when he needed me, thinking he would give me a week or so to crawfish out of it, and he said, "Tonight, at 5 p.m." The heat was on.

I then inquired what to bring, and he responded laconically, "Plenty of warm clothes." It was October after all. I almost asked, "Are they nice guys?" but I thankfully didn't. I did picture myself lifting some heavy stuff, which turned out to be the only accurate component of the image my mind had dreamed up. The part of harmonizing to "Old Man River" didn't ever quite turn out. Like the legendary Robert Johnson, I had brought myself to the crossroads. Would the devil be waiting for me there to steal my soul?

I will say this for my dad. Having spent years on a small ship in the Pacific during the great war, he tried to talk me out of going. But, I had at least half a mind to do this, so I asked him to drive me to the river that evening. I lugged my onion sack on down to the loading spout where I met two dock workers who shook their heads and chuckled when I claimed the boat would be there at 5 p.m. They sagely advised me to sit a spell and relax with them. We smoked and talked about a number of things before one of them got directly to the point and asked, "Man, you got a girl?"

I returned a perfunctory nod because I sort of did, to which he said, "She ain't gonna wait no month for your ass to git home. From now on, man, you just gonna have to catch as catch can." I laughed along, but I feared that his joking words could one day come to be a prophetic statement. Concerning the next eleven years of my life, they were at least relevant.

At seven o'clock, the boat still nowhere in sight, my dad left the nearby tavern he had chosen to wait in and came back. We rode around in his Chevy, with a three on the tree, as he tried continuously to get me to change my mind. The Cincinnati Reds were playing the Oakland A's in the World Series, and we listened to the game on KMOX as he drove. After about an hour, he took me back to the dock, perhaps hoping I would grab my duffel and leave. Still, no boat had arrived. We repeated this cycle a couple more times, and I found myself gradually starting to lose some of my resolve to sail the seas dry. At 10:30, I was about to give in and go on back home forever when I saw a searchlight coming up the river. This had to be my towboat! I shook hands with Dad, wished him luck, and told him to be careful going home. I stuck a Lucky Strike in my mouth to look bad, prayed that the bagpipes would not call me home before my thirty days were over, and climbed a ladder down into the mystery below. I'd be better off, I thought, to keep going steadily forward, or I might take a notion to end this fantasy before it ever got started.

As my steel-toed boots—that the company had told me I needed to have—hit the deck, a guy close to my age met me, sized me up, and asked me what part of Meade or Breckenridge County I was from. Bewildered at that greeting, I asked, "Where's that?"

Equally flummoxed, he spat on the deck and said, "Damn." All the other deckhands on the boat, I soon discovered, were from one of those two counties in Kentucky. A beer joint there, known as Dead Horse Holler, was utilized as an ersatz hiring hall. It helped your cause if you also came to The Holler with hogs to sell. I had immediately been identified as an outlier, one with no hogs.

My new buddy led me through a door into a smoky lounge filled with outlet furniture and peeling floors, then to what he told me was to be my room. Without knocking, he burst in, turned on the light, and startled my new roommate, Bobby, who was asleep in the upper bunk. Bobby leapt from the bed as if the entire Commonwealth of Kentucky State Patrol was suddenly in hot pursuit and cut his foot open on something when he landed on the deck. He proceeded to bleed and cuss with a relish I had never before heard, and I was hoping my dad maybe hadn't left yet. But Dad was gone. My last lifeline to home had pulled off, and I was soon to be alone in a dark, all-too-cozy room.

As Bobby continued to shed blood all over his Green Lantern comic and carry on the string of astonishingly creative obscenities, Jace, the first guy I met, told me I was lucky because I was on the front watch and was just getting off duty. I would begin my river career by sleeping, or trying to, after Bobby was patched up and the blood was more or less cleaned from the floor.

After they left the room, I took a few clothes off and turned out the light. As I lay there, rolling over the sheets in the enveloping darkness, I suddenly missed my mom and dad and longed to pet my little brown dog again. Soon, the beating and banging started. It would ease up for a spell, then emerge again unmercifully. I felt as though the river would come into my room and drown me in my bunk at any moment, sorrowfully creating an inglorious ending to a brief river career. I didn't sleep that night or for the next five days.

Somebody turned the light on in my room at an early hour the following morning and told me it was time for breakfast, even though it was still dark. I went into the galley and greeted the cook, an older lady, and surmised that I should be extra polite to her. She asked me how I wanted my eggs, and I figured scrambled wouldn't put her out too much. I went to the table where everyone was sitting, and they all snickered in anticipation as I grabbed the only available chair—at the end of the table. A few minutes later, the pilot, who had gotten off watch, came down into the galley, tapped me not too gently on the shoulder, and said without amusement, "Whachu doin' in my chair, boy?" That had to be the highlight of the week for everyone. They all laughed, and I figured it was best politically to laugh along with them, after I dispatched myself quickly from his chair.

After breakfast, the cook handed me an egg-and-bacon sandwich and an apple and told me to take it up to the captain. When I finally found my way up to the pilothouse, day was just breaking, and I could see we were in a town, an industrial one with high smokestacks. They had told me we were heading up the Illinois River, and I figured we had surely made it to Springfield or Decatur or somewhere up in Illinois in six noisy, frightening hours. I gave the captain his sandwich and meekly, speaking over Loretta Lynn, inquired what town it was we were in.

He turned deliberately around in his chair and stared back at and through me with a frown that advertised, "If you ain't a dumb son of a bitch, you'll do just fine until a real one comes along." After a few awkward seconds, not knowing whether to scratch my ass or do jumping jacks, he deigned to speak in my general direction. "You from Sa'nt Louis, ain't ya, boy?"

Somewhat honored that I was at least recognized as being alive, I said, "Yessir."

"You don't know your hometown when ya see it go past ya?"

I shook my head pitifully, like the dumb motherfucker I had now unmistakably been pegged as, and silently assumed it might be best to not ask any more questions for a day or two, maybe for a lifetime. I was also contemplating about what an arcane means of transportation to the adventurous side of life that I had found myself a part of. After six hours of bedlam and collision, we hadn't even made it out of town!

He told me that the mate, Swede, would be my direct supervisor. Swede was sleeping after a tough night but would wake up later and explain my duties to me. In the meantime, I would follow Jace around and pick up what knowledge I could. I learned a little bit about Kentucky that morning, like the fact that I was also a dumb motherfucker because I didn't know it got down to minus ten in the winter there.

The chief engineer told me that if I ever wore a pair of boots out on the deck, the river would be in my blood forever. He also said that those who start in the winter, if they live through it, will be tough enough, or dumb enough, to stick it out. These aphorisms came out as comforting, sagaciously lyrical poetry. Things had brightened up a bit. The chief liked to stay in the galley and drink coffee with the cook while considering other appropriate platitudes.

I proceeded to learn all kinds of other invaluable information, like there is no such thing as lunch. The noon meal is called "dinner," and the evening repast is "supper." Swede got up at dinner, and that's when I started to learn what a d.m.f. I really was and that the phrase would become my unsolicited sobriquet for the near future.

Well, one thing I was to find that I bonded with my new family on was that we were all much happier when the mate was in bed. He came to dinner in a filthy Economy Boat Store T-shirt that had the stomach ripped out of it. We were all thusly treated to an archipelago of hair sprouting freely from around his navel. His trousers could have stood in a corner by themselves, and he wore no socks inside his tattered canvas loafers. Where were his steel-toed boots? He made snide remarks at the table concerning the food that I thought were uncalled for because the vittles were quite good.

After dinner, Swede disappeared, and I tried again, in futility, to sleep. There was only one communal bathroom on the first deck at the end of the hall, and at about 4 p.m., I felt the need to leave my bunk to urinate. As I reached the urinal, I could see Swede's faded canvas loafers sticking out of the closed commode area to my side. I could hear him turn the pages of a magazine and grunt luridly as I relieved myself. Spending the next half century of life on the river was certainly not on my mind in the midst of that glorious moment.

After supper, we had stopped for some reason that I didn't understand, and I started to play a harmonica I had brought with me to kill time. The captain must have seen me and ordered Swede to get me busy doing something to earn my twenty-five dollars a day because he petulantly hobbled down from the pilothouse and told me I had to learn how to toss a line. Of all the things I learned on the river, tossing a line turned out to be my favorite. It made me feel stunningly folkloric, like Roy Rogers or Gene Autry. I learned that cowboys use ropes, while rivermen use lines; as Swede said, "You ain't worth a shit if you can't lasso a timberhead." I at least wanted to be worth a shit, so while the other guys were smoking, drinking coffee, complaining about the food, or just hiding from the mate, I was out in plain view of the captain learning how to throw a line. It never really got me anywhere to speak of, but it kept me occupied.

I did learn that the reason we stopped was for lock delay (waiting your turn to go through a lock). Since it requires minimum effort, that has turned out to be something I excel at. We waited at old Lock 26 in Alton for up to three days at times, and I was steadily learning to throw lines. Some of the others joined in, and we had a grand old time making bets on tossing lines and listening to "Summer Breeze" and "Everybody Plays the Fool," among other things, on an old portable radio one of us had. There were other responsibilities as well, such as cleaning the pilothouse, making the captain's room up, burning our trash in a "burn barrel" on the stern of the tow, and cleaning the stove after the cook had trashed it out all day long.

The big event, though, was tow work, which consisted of taking individual barges and lacing them together in a steerable conglomeration of metal. The mate hated tow work. Therefore, he made it hard on the rest of us d.m.f.'s. When we were climbing around on those barges and lifting ratchets (that felt like Refrigerator Perry was hanging on the other end), wires (or "wahrs" as most of the boys called them), and slings, we were all d.m.f.'s in the mate's not-too-modest opinion. This analysis included the captain when he took too long to slide a barge in tow. I must admit that I was the most hapless member of the deckhand cache, and Swede would curse, barely under his breath, about d.m.f., "green" deckhands that the office was sending out. I felt uniquely special knowing that I then was color coded, to go along with everything else.

After a couple of weeks or so, I got to the point of being able to sleep for a few hours, the boys from Kentucky joked sociably with me some, and I had at least learned to tighten a "wahr" without getting the toothpicks (ten-pound metal batons used to tighten wires with) fouled. One day, I had somehow gotten left on watch by myself, and the mate was mercifully sleeping, I hoped like Rip Van Winkle. The pilot rang down in the galley for me to come up to the pilothouse. I walked in, and he said, quite nonchalantly, "Boy, git yor spike pole out an' fish that deadman outta the river. I know yo're by yorself, but a good deckhand kin pull a deadman out by hisself."

He said *what*? There's a dead body floating in the river out there—and I have to pull it out? I stood in stunned disbelief for close to a minute thinking, "How often does this event occur?"

"You hear what I said, boy?"

"Shouldn't we call the police first?" I offered in weak response.

Then the cognitive turmoil became his, as if he had unexpectedly been informed that premium gas for his 442 would someday cost him a buck fifty a gallon. Left momentarily speechless, he analyzed the depths of my ignorance. Finally, he spat words out in utter exasperation.

"Boy, a goddamn deadman is a fuckin' shore wahr. Git that fuckin' mate up ta larn ya somethin'!"

So, a deadman is not a waterlogged homicide victim, rather a shore cable anchored into the bank and used to secure barges with. Imagine that.

A couple of weeks later, I had completed my first hitch and was relieved by another green hand. I went home and told everyone I knew about all of the wonderful lessons I had learned on a towboat. In a few months, I wore out my first pair of boots. Fifty years later, I am still learning things and somehow muddling along through other "imagine thats." The strong, brown god has an endless supply of them.

# CHAPTER 3

# Deadly Decks

I do not know, statistically, how a deckhand's job rates, danger-wise, with that of an ironworker, oil-field hand, firefighter, or public-school teacher in a big city, to name just a few scary professions. However, I do know that there are several law firms in St. Louis, Cincinnati, New Orleans, and other river cities that specialize in admiralty law, and over fifty percent of their cases deal with personal injury.

I took the big plunge into Mama Mississippi three times in my decking days. According to river-safety specialist Dennis Kundert of TECO Transportation in Devant, Louisiana, "man overboard" still ranks as the biggest hazard faced by a deckhand. Kundert told me that 95 percent of all man-overboard accidents are caused by carelessness and neglect rather than unexpected or unpreventable circumstances. Luckily for me, I always fell in with a wire or line firmly in my grasp and in the presence of my "deck-dog partners" who eventually got around to helping pull me out, after they had their chuckles.

Man overboard is, of course, serious business and is a required drill for all boats. Even if one is a good swimmer and is wearing a life jacket, falling in the river can often be fatal, especially at night. If an incident occurs while the boat is underway, a person can be sucked under the tow or into the wheels (propellers) of the towboat. Winter, naturally, brings on the added hazards of icy decks and hypothermia. Anyone immersed in forty-degree water will begin to lose coordination and strength in less than a minute, much less time than it takes even a well-trained deck crew to launch a rescue boat.

Although pilots, engineers, and cooks seem to be less exposed to the risk of personal injury, they sometimes end up in harm's way as well. Poor Mick, an engineer I often towboated with and who took a lot of our money playing poker, fell in the river in St. Louis at about 11 p.m. one November night. At about 11:20, he was still clinging to the guard chain he had managed to grab before going in, when some crew members happened to hear his weakening screams. In an unselfish act of Christian charity, they pulled him out without first asking for some of their losses back. Many large towing companies now require anyone going outside on the lower deck to wear a life jacket and to be accompanied by another crew member.

Although it's the most common and fatal hazard, man overboard is not the only killer or maimer of deckhands. The rigging and lines employed to hold barges together or to lock walls must always be respected, inspected, and quickly rejected when they become too hot to handle. Those wire cables and two-to-three-inch-thick synthetic lines and ropes, while capable of absorbing tremendous amounts of weight and energy, can also mercilessly crush fingers, sever arms and legs, or decapitate the untrained or unwary deckhand.

A very sobering picture was on display at Lock 20 at Canton, Missouri, when I was a deckhand in the 1970s. In the photo was a broken lock line and a pair of work boots, lying empty and ripped apart on the deck of a barge. Those boots had been on the feet of a deckhand until the fury of the line had knocked him completely out of them and cruelly ended the young man's life. Perhaps the Corps of Engineers thought the image a bit too grim because the photo is no longer on display at Lock 20, but it is still vivid in my mind.

Less dramatic and menacing, though still costly in terms of pain and man-hours lost, are the sprains, strains, and contusions associated with a deckhand's job. Deckhands are constantly exposed to cold, heat, wind, darkness, and the numerous forms of precipitation that dwellers of the Upper Mississippi Valley have grown to adore. When there are barges to be wired together or a "lock on the head," the show must go on, regardless of conditions. The rigging and lines are cumbersome and heavy. Ratchets and wires can weigh as much as a hundred pounds or more, and the deckhand must learn to sling them up, over, and around deck fittings, barge coamings, and hatch covers. All this has to be done in poor visibility, with water streaming off the deckhand's cap and out of his nose, with wet socks falling down into his boots, which are often ankle-deep in black coal or mashed, rotten soybean muck.

Keep in mind, also, that most crew members work thirty straight days or more at a time, and few have had more than five consecutive hours of sleep for that entire period since they work six hours on and six off. Particularly onerous is the position of "call man" that most companies now employ. On the Upper Mississippi, this lucky guy gets to experience all the fun because he has to get up for every lock and landing from St. Paul to St. Louis, that is twenty-six locks and who-knows-how-many landings. The call man will often get less than an hour of sleep between locks or landings and is indeed fortunate to have more than four or five uninterrupted hours when the boat is above the Quad Cities.

Despite the questionable practice of decreasing crew size, thereby increasing individual workloads, towing companies have greatly reduced job-related injuries and fatalities beginning in the 1980s, according to Kundert. Most companies have come to value the economic importance of safety and health on the river.

Indeed, I was fortunate to make it through my first six months as a deckhand unscathed as I often went alone on the tow over icy decks with barges gapping more than a foot apart. Leaping from one barge to another was just an accepted risk that had to be taken. Though official company policy was to don life jackets when on the tow, in practice, most captains looked the other way. It was a culture of risk taking.

Today, most green deckhands attend a training school before being sent out to a boat. Some companies go the further step of "riding heavy," which means the new man rides as an extra for a week to get an experiential taste of decking under the watchful eye of a mentor. The mate, the foreman of the deck crew, is now trained in safety and first aid and conducts periodic safety meetings with the crew. Many companies offer incentives and bonuses for safe work and have taken on the expense of hiring a safety manager or consultant.

Despite this improved focus on safety, many of the inherent dangers of working on the river remain from the days of log rafts, steamboats, and keelboats. Perhaps, it is the desire to take chances or touch the unknown and mysterious darkness of river life that draws people. If you want, let it draw you. But have your life jacket buckled, your steel-toed boots on, and your flashlight in hand, buddy! And don't go out on the tow alone! It's dangerous out there.

### Author's note

I wrote this article almost thirty years ago. I had gotten Mr. Kundert's name from a friend who worked for that company. Mr. Kundert was kind enough to provide me with a good amount of information for this piece over the phone. I never met him in person and am unsure regarding his current affiliation.

# CHAPTER 4

# Splicin' Line

His hands moved purposefully and so swiftly that you could not notice that two digits were missing on the left side. An unfiltered Camel hung from his lips, and the frequency of their use had left eight brown-stained fingernails and a throaty cough. The sleeves of his blue work shirt were rolled to the middle of forearms whose veins bulged as the vicelike fingers twisted and tucked the line with rhythmical dexterity. Beads of sweat patiently queued and awaited their turn to drop from his chin, darkened by years of exposure to sun and wind. His given name was Eldred Simms, but nobody ever called him that—to us, he was "Pappy."

We had fifteen loads of shiny, black, southern Illinois coal heading up the river to Wisconsin that morning, and our job was to splice eyes in that new coil of lock line we had picked up at Economy Boat Store the night before. Actually, I should say that Pappy's job was to splice the line—and see if he could teach me. The decks still retained their morning dew as we stretched out six hundred feet of thick, white line in the cleanest, driest area we could find.

"This here tuck is the hardest," Pappy began. "Ya gotta turn the whole thing 'round ta the right. Now, see, it look like she's a-comin' outta here, but really she come up through thar!" His emerald eyes twinkled as he divulged this mystery of line splicing to the scared, homesick, and confused student that I was.

"That mate's bin kinda hard on ya, ain't he?" he said, stabbing at the crux of my misery yet not moving his glance from the job in his hands. I was too ashamed to say "yes" and too honest to say "no," but Pappy recognized my silence as consent. "Some of 'em thinks they owe it to a new fella ta make it hard on 'im, some of 'em's just a-tryin' ta give out what they hadda take. Don't let it git ta ya."

"You ever thought of runnin' mate, Pappy?" I ventured.

"Oh, I already done that—now, after this un, ya jist take an' start it all over agin." He stood over six feet and weighed every bit of two hundred pounds, his voice melodious—pure and deep, as though it came up from the river itself. "Jist repeatin', that's all line splicin' 'mounts to—I don't keer 'bout no mate's job no more. I leave it to them young uns wantin' ta git 'em a pilothouse job. They'll crawl all over each other ta git up there."

"You ever wanted to be a pilot?" I asked as we passed Grafton and the Illinois River.

He didn't answer but looked away from the line and out over the river for several seconds. "Yeah, I did. Woulda did anythin' ta be a pilot, an' they knowed it," he said, nodding up at the wheelhouse that towered stately above us. "So, a cap'n, he kep' me out six hours in Peoria Lake one winter when it was thirty below an' we was a-mule-trainin'. My gloves got so froze, they warn't no good ta me no more—so I los' the feelin' in my fingers an', a coupla days later, I was an eight-fingered fella," he said with no discernible bitterness.

"Did the company take care of you?"

"If that's what ya wanna call it. Paid fer the sawbones, but d'reckly I was back out on a boat. A man couldn't afford ta stay off in them days—somebody'd git yor job. It only hurt on cold days fer a coupla years.

"Here, remember how we do this las' one? Point ta me where it goes."

I took my best, educated guess, pointing uncertainly to where I thought was the correct strand of the line. He took the Camel out from between his lips and spat out a few brown weeds on the deck. His face broadened as his teeth emerged from where the cigarette had been.

"Yo're ketchin' on," he smiled. "Now. This here line," he continued, "it's a plastic, son. Poly-d they calls it. If'n ya think it's a-gonna break, ya run like hell away from it. Saw a boy git knocked plumb outta his shoes by a broke line one night."

"Did it kill him?"

"Ya ever seed a body alive what got knocked outta his shoes?" He paused briefly, then continued, "Now, one more setta tucks an' we'll be done with this un. Then I'll let ya do the nex' un."

"Good—if I can learn to splice line and make bumpers, they'll give me experienced pay," I said with a degree of eagerness.

For the first time, his eyes faced right up to mine.

"Son, when ya larn somethin', don't larn it 'cause somebody's gonna give ya somethin' fer knowin' it. Know it fer its own sake. Cain't nobody take it 'way from ya then. 'Cause it'll be part o' ya.

"Now. If'n we got us any loose ends, here's where we cuts 'em off."

He took out a sheath knife, instructed me to hold a group of loose fibers taut and, with his forearm completely relaxed, let the well-honed blade do its task.

Nobody messed with Pappy. He was the oldest man in the crew, but the respect paid him went beyond mere deference to age. It was a primitive, savage tribute akin to what a pack of wild animals have for the one whose fury has been witnessed or reputed and is so decisive and clear that they wish to never experience it again. The mate, whose tongue could lash at you like a bloody whip, merely consulted quietly with Pappy when he wanted a task completed, and Pappy never failed to dispatch it with exquisite brevity and efficiency. Once, I saw the mate, two deckhands, and an engineer curse and cajole an outboard engine that refused to start. In the middle of this slurry of words, I heard the gentle sound of oars dipping in the water as Pappy, fatigued from the delay, rowed a smaller yawl to shore to finish the job. Pappy's power, however, was the ultimate trump card that was never led, but saved for all to know and respect.

"How long you been on the river, Pappy?" It seemed a logical question, for want of anything better.

"'Bout halfway 'tween Helena and Greenville, they's some shanty boats on the river. I 'as born in one of 'em. So, ya might say I bin on it my whole life." Then he stopped and looked at me again. In his gaze there was an uncertainty that I had not before seen from him, as if he were standing on the edge of a platform wanting to take the first step, to reach out, but unsure of his trust in the venture. I had only known Pappy for a short time and our conversations had been friendly but stunted, and I had gathered that this was his way. I expected he would go no further, but . . . .

"My daddy, he was whatcha call a lamplighter—kep' up three navigation lights.

"Here, this un's done. Now ya cut the end o' this un. Make sure ya cut it clean.

"Ev'ry day o' the year, my daddy would go out an' check them lights and, when I come of age, I'd go out with 'im. We'd fill 'em up with kerosene, light 'em, an' cut the weeds from 'round 'em.

"Here, don't use that there axe fer no batterin' ram, use it gentle-like, let the blade do the work fer ya.

"It was hard work and Daddy, he was a hard man." He dragged a steel-toed work boot along the barge, his eyes riveted to the deck. His voice softened and slowed.

"'Course, mebbe I cain't blame 'im too much. It was lonesome fer him out there after Mama lef'. And he was a drinker. I reckon he jist had somethin' inside o' him that hadda find a way out. So, sometimes he was mean an', sometimes—well, he was bigger 'n me." His words began to choke off.

"Looky heah, now. Take that line, son, an' separate them ends down—that's it! Not too far, now, an' then serve them ends.

"But, he did larn me the river, my daddy, larned me how ta row a boat, 'bout lines an' such. He was like the river. When he warn't a-drinkin' an' when we was out there a-tendin' them lights, he was peaceful and calm; jist like water on a July evenin' when that sun seems like it's a-gonna stick right above them trees all night 'til the sky and it turns red together. But, sometimes, he was like a storm in January when it'd all well up inside o' him an' them waves'd git bigger an' bigger, 'til they hadda 'splode—an' they was cold an' bitter. Tha's when I larned how ta duck 'im an' I swore that soon as I's old enough, I'd hafta leave—or kill 'im. It was down ta that.

"Now, ya kin start yor splicin'," he continued. "Yo're a ten-fingered fella, so's it oughtta be easy fer ya. But ya need this here fid, anyways, 'til ya git that stren'th in yor hands."

"What'd ya do, Pappy?"

"Oh, I run't off. When I's 'bout fourteen."

"Where'd ya go?"

"Well, they was lots o' fellas a-travelin' 'bout in them days. I didn't have no money, so I'd hafta luck inta jobs. Mainly, I wanted ta git as fer a piece from my daddy an' the river as I could git—ta wash it away, fergit about it. I did know how ta work hard, an' long, without eatin' much an' I warn't 'fraid o' no cold. Then, one day, I comes to this place out in Nebraska, an' it don't seem like no big river was 'round there."

"Are these ends 'bout long enough?" I asked, displaying what I had cut.

"That's jist 'bout right, son. Now, lay it over so it's all nice an' purty-like."

"Don't matter if it's pretty or not, does it?"

"Course it does! It don't affect the stren'th none, but there's gotta be some purtiness in ev'rythin'. Remember, that splice is part o' ya—says somethin' 'bout ya. That's a Lee splice! Ya wanna be proud of it, so's ya kin stick yer chest out! Now, roll that fid. It'll be a sight easier on ya. Where was I?"

"You was in Nebraska."

"Yeah. Hired on with a man—a rancher. He didn't have no son lef'—hadda wife, and a coupla daughters," he said. "Well, I worked hard fer him an' I reckon he figured I was 'bout the age of what his son woulda bin if'n he'd a-lived.

"It was a home like I never had before. It warn't like nuthin' I'd ever knowed. I had me a family. An' I seen they's another side ta life." He paused and stared into the sky. "But, they was somethin' inside o' me. Jist like that storm inside my daddy. I reckon it was Daddy's blood—though I might try an' deny it an' try an' fight it away, it still runs through my veins. So, the war come, an' I lef' ta go—an' when it 'as over, I couldn't go back there."

"Couldn't?"

"Couldn't, son." He looked over the corner of the barge we were sitting on and threw his Camel into the river. I knew that this chapter was closed. "Jist keep repeatin' them tucks. Like ya bin a-doin'!" He said crisply.

He didn't speak again for what seemed like several minutes. My thoughts briefly turned inward, then back to him. He had opened the gate up briefly, and his words had flowed like a mill race, and just as suddenly, the sluice had closed. I was content with working on my line and listening to the sounds of the water breaking and gurgling around the corner of the barges. The gulls screeched as they darted and dodged one another in the heavy morning air.

"So, then I come back ta the river," he finally began again. "It was like a pull that I couldn't turn down. Once ya've wore out a pair o' boots on this river, it'll grab ya fer life. Ya done heared that un before. No matter where ya might go, when ya close yor eyes, ya'll see that river. That brown water'll seep inside ya, an' no place'll seem more like home."

"How 'bout Rex?" I interrupted. "He says this is his last trip."

"Ol' Rex bin a-sayin' that since I knowed 'im, an' that's bin eight year.

"Hey, yor line's looking real purty thar!

"So, I got me a job outta that hirin' hall in St. Louie. Started in December, jist as ice season was a-comin' on, an' the mate, he says ta me that I'll be out here a long time 'cause a fella who starts in the winter has gotta be tough enough, or fool enough, ta make it. But, hell, I knowed I was gonna stay anyhow.

"An' then one day we goes by Terrene Landin' an' I seen Daddy's ol' shanty. So, when I gits off the boat, I catches me a bus to Rosedale—but I jist couldn't bring myself ta go an' see 'im. I heared he's still 'live today—jist too damn ornery ta die, I reckon." He lit another Camel and moved across the barge to a timberhead and sat on it, looking into the water.

"But, we was in a slow spell an' I had me a coupla month off, an' while I was in Rosedale, I met me a young gal an', when I goes back ta the boat, I had me a ring on my finger, an' a baby on the way. We didn't know what it was gonna be like, me an' her—ta be married an' be hunnerds o' miles 'way from each other. She was so young! Hadn't never even bin outta Mississipp'. Son, no matter how tough it is out here fer ya, it's three times worst fer a woman on the bank a-waitin' fer ya ta come home. We didn't work no day fer day back then. They 'spected ya'd work sixty days an' only take a week ta ten days off. Yor home was only a place ya'd go ta visit. So, while I 'as on the boat, I 'as like a tiger in a cage—I had this big ocean wave inside o' me, pent up an' jist a-waitin' ta crash, ta 'splode. I'd git off the boat an' go ta her and love her. But then, a coupla days after I'd bin home, that wave'd be a-lookin' fer someplace else ta go, an' it jist didn't have no other good di-rection. I knowed it was part o' my daddy in me, but I couldn't let loose of it.

"Sometime, I'd try an' tell her what I was a-feelin', but it didn't work. I'd go back ta the boat an' buy her somethin' purty 'fore I lef'. I was empty in my guts from leavin' but didn't have no other job ta go to, an' that river was a-callin' ta me, like some long, tall, painted woman with a red dress an' her arms open ta me—an' that wave knowed how ta go thattaway. It was easy.

"I'd try an' write to 'er, but, shoot, I couldn't hardly spell nuthin'. She'd write back an' I'd git somebody ta read a letter fer me an' it felt good—like, this time when I got off things was a-gonna be diff'rent. The baby come, she hadda tough time—I heared. I was on the boat when it happened." He looked down between his feet, and his voice dropped to a whisper. "Shoot, Lee, warn't even there when my own son got born. Them letters got futher an' futher apart, an' shorter. An' then, I gits off the boat one time—an' she was gone."

I uttered a curse.

"I said that, too, an' worst, an' swore I'd find her an' git her back. But—I never done it, an' after a spell, that quit a-hurtin' too. She done what she hadda do. I don't hold nuthin' agin' her. She's raised that son o' mine real good with what she had. He's in college, ya know, a-studyin' ta be a lawyer. Hell, I ain't a-seed 'im since he was a baby. Wonder if she's tol' him his daddy's an ol' river rat?" He smiled and stared vacantly in my direction, then said, "One more setta tucks an' ya'll have it whooped, Lee-Lee!"

In the period of silence that followed, I worked furiously to finish my splice. I slowly lifted it up for Pappy's inspection, wanting him to be proud.

He took it gently in his hands, inspecting every tuck, his eyes not missing a thing. It seemed an eternity before he softly said, "That's a fine job, son. Yo're gonna make a river man."

If Mozart would have been alive to tell me I would write a symphony or if Babe Ruth would have called me "slugger," I don't think I could have been prouder than I was at that moment. I wanted to take my line and show it to the world. It was mine—and it was beautiful!

In the midst of my revelry, I looked at my teacher. He sat on that timberhead, silently powerful, like a mountain in morning sunlight. He was silhouetted with the chalky bluffs of Illinois to his back and between, the river—flowing. The river that had taken so much from him but that he could not leave because it defined who he was.

"Pappy, when we get up to Lock 24, we'll be off watch. Can I buy you a beer?"

"Naw," he said. "Ya don't owe me nuthin' fer that. 'Sides, I quit a-drinkin.'"

"Well, I just thought . . . I mean I wanted to—"

"There is somethin' ya kin do fer me."

"What is it?"

"Ya kin take what I larned ya today an' larn it ta somebody else."

It seemed so trivial a request then that I quickly, almost thoughtlessly, made it. Since that day, however, I have come to realize the depth of that promise I made. I have attempted to take that torch from him and done my best to pass it to others. No, there isn't any book that you can buy that will teach you how to splice a line or to stop a first cut or to carry a ratchet or to ride the head of the tow through an icy rain or to sit in the warmth of a sunset or to cry when the river has taken everything from you. You can only get it from someone who is willing to let you touch it and let you know how it touched them.

I want to tell you a few more things about Pappy that I have since learned. He was remarried when I knew him. He and his wife were too old to have children, but they had adopted two who had been abandoned. I think they might be two of the luckiest kids in the world. He is home six months of the year now, and though that wave still lives inside him, it has become a warm, gentle one that brings peace and knows its way home.

I hope that Pappy's son has become a good lawyer and is rich and prosperous. I also hope he realizes how fortunate he is to have Eldred Simms's blood in his veins—to have part of his soul. If he doesn't know it, I'd sure like to tell him.

It's been almost fifty years since that day above Alton. We used my line on the head of the tow at Lock 25, and when my spliced eye held the weight of 30,000 tons of coal and steel, I wanted to scream it out to the world that it was mine. You can't carry a lock line with you in your hip pocket, so I can't say that I still keep it as a remembrance. The line doubtlessly met the fate of all other lock lines: being used, abused, finally broken, and made into bumpers or lost in the river. But I still have that memory of it, and it seems like just an hour or so ago that Pappy larned me how to splice a line.

# CHAPTER 5

# A Warm Cup of Coffee

The following is a story that I have heard so many times from so many different rivermen that I know there cannot possibly be a single word of truth in it. Still, it does have some worth in terms of showing how many rivermen feel about a certain issue, and that is unionization of labor.

Strangely, organized labor movements have not always done well in the river industry, especially with marine officers. I say that this is strange because rivermen, since the nineteenth century, have been grossly underpaid by employers. Though Mark Twain writes of antebellum pilots making the astounding sum of $300 per month—in those days, equivalent to the earnings of most doctors, lawyers, or bankers—the current pay of pilots is usually not at all reflective of the demands of their jobs.

When I was a pilot in the late 1970s, I was paid the exorbitant sum of $190 per day. I thought, at the time, that I was very well paid and was, in relation to what I had made before, until I put it all in perspective. It is a matter of fact that an ocean captain was paid at a rate of about two and a half times that of a captain on a thirty-five-barge tow, and a very knowledgeable Coast Guard lieutenant once told me that he thought towboat pilots are arguably the "best hands-on navigators in the world."

The main reason that river folk's wages are so low is their inability to organize. There was a period, however, in the 1960s and '70s when the unions were able to gain footholds in the towing industry. The National Maritime Union and the Seamen's International Union were able to raise deckhands' wages and to specify certain tasks that they did not have to perform. Previously, deckhands were pretty much expected to do anything that an officer instructed, hoping that their will was at least benevolent.

It had always been a river tradition for the deck crew to bring coffee, tea, snacks, etc. to the pilot on watch upon his request. During contract negotiations in a certain year, however, the union had managed to sneak in a proviso that the deck crew was no longer obligated to serve the wheelman in this fashion. Their duties were to consist of boat maintenance and tow work (locks, fleet work, tying off the tow, etc.)—and they carried a brand spanking new copy of the contract on every boat just in case there were any questions.

So, this story goes that one rainy night on the Lower Mississippi, on one of the graveyard reaches between Baton Rouge and Natchez, a lonely pilot in a cold pilothouse called down to the warm galley on the loudspeaker to request that a deckhand bring him up a hot cup of coffee. There was much shuffling of chairs, harrumphing, and pointed conversation down below, but when one of the deckhands found his way up to the pilothouse a few minutes later, it was without the expected coffee.

He proceeded to plop himself authoritatively on the liar's bench behind the pilot. After a few moments of awkward silence, he blurted, "Cap'n, it ain't in our union contract to bring no coffee to no pilot."

The pilot contemplated the situation silently for several minutes before asking, "Is it in your contract to tie off the tow?"

"Yep, we still do that and tow work," came the reply, steadily gaining in self-perceived credibility.

"All right." And the pilot said no more.

The deckhand, union contract clutched like a medieval mace, was disappointed that the issue had not been further debated, boredom sometimes being a problem northbound on the Lower. After a spell, in which he must have assumed a stance of verbal victory, the deckhand haughtily left the pilothouse. But this story was just beginning.

It was a few minutes later as the deck crew celebrated over a game of rummy that another call came over the loudspeaker in the galley. *"Let's find a tree to tie off to"*—something still quite specifically in the union contract.

The three deckhands donned gear and stumbled out through the now rainy, cold night air, dodging deck fittings and rigging with their often-malfunctioning, aluminum flashlights a quarter mile out to the head of the tow only to find that the only line available was not long enough to reach the usable timber. They then had to walk back to the boat, pull a two-hundred-foot lock line from the deck locker, divide it in three sections, wrap it like a huge python around their shoulders, and lug it that same quarter mile in the intensifying rain.

Bending the two lines together, a couple of the brave souls had to climb down a shaky ladder, wade through calf-deep water, and pull the line up over the bank, through fields of poison ivy, around a tangled cottonwood, slipping in the mud, cursing at each other and the world. After encircling the tree, they pulled the eye of the heavily soaked line back up the ladder with them to the empty barges that sat twelve feet straight above the river's edge. Finally, having secured the tow, bruised, disgusted, and dripping nasty stuff from their bodies, the proudly unionized deck crew slogged back to the boat after the hour-long bonding experience.

They arrived back at the boat and headed to the galley and found the pilot, who had come below to make his own coffee.

Smiling over a steamy cup that had been made personally to his own taste, the pilot calmly said, "Okay, you can go back out and turn me loose now. The weather's cleared up, and I made my own coffee."

And so the story goes, the next time the pilot called below for coffee, that deck crew was scrambling all over each other to reach the pot first.

### Author's note

I wrote this story in the 1990s. Unions on the river, for the most part, have gone the way of the dinosaur. However, wages on the river have improved so much since about 2010 that some septuagenarian pilots, like this writer, are trying to make up for the years of depressed earnings, well after retirement age.

Though most towboat pilots are not union members, I have been a proud member of the American Maritime Officers (AMO) and am still drawing a pension with them. I was with AMO as a pilot for the Delta Queen Company. I was also a member of the Masters, Mates, and Pilots (MMP) while with the Corps of Engineers and valued that professional partnership.

# CHAPTER 6

# An Insignificant Participant in River History

In case it never occurred to the reader, towboaters don't take a holiday break. For a warm and fuzzy Christmas feeling, you'd be better off hanging with Scrooge and the Ghost of Christmas Future than being on a towboat. There are few Christmas trees, therefore fewer presents underneath, no lights, no carolers, and no Santa with red-nosed reindeer. Shepherding those barges safely up and down the river is the mission, just like it is on any other day.

My first Christmas Eve on the river was a white one as I went out to tighten rigging and earn my twenty-five dollars per day. Carbon searchlight arcs broke through the snow like light sabers, highlighting individual flakes. They were falling fast, some melting on contact with barge steel, while others began to accumulate on my hat and gloves. Though they brushed off easily, the white powder gave notice that it would continue in earnest for several hours. From houses perched on the bluffs of the Illinois shore on Alton Lake, decorative lights could be seen aglow, framing kitchens and living rooms. Somewhere up there, I was certain, someone was sitting beside the fire with their black Lab. My mind, as it has been for half a hundred years, was torn between a longing to return to the insulated comfort of a home and hearth or to remain in the isolated orbit of the river traveler.

The snow and clouds held some heat in over our tow, and I began to sweat inside the coveralls my mom had purchased for me. Mom never wanted her son to experience discomfort and, I am sure, was sleepless with worry over me. No matter, armed with my toothpicks and cheater pipe, I flexed muscles on every wire on the tow. Wires need to be tight for the pilot to steer the phalanx of barges, and I was determined to get them tauter than a bow-fiddle string. Tightening rigging, in my eyes, was better than cleaning up the galley, which I would also have to do later. Maybe the next day, I would quit the river and go home for good—but probably not. Since 1972, I have entertained those thoughts on a daily basis.

As we passed by the gaudy Piasa Bird figure lit up in the quarry near Norman's Landing, I remembered from high school history class that Jacques Marquette and Louis Joliet had canoed these waters three centuries before. I was, at least momentarily, inspired to regard myself as a partner in their legacy. Though not a very illustrious partner at the time, I was still following in the footprints of all the other rivermen who had preceded me.

To be sure, Native Americans had used the rivers for trade, travel, migration, and war long before the Spanish Conquistador Hernando de Soto first laid his depraved eyes on the Mississippi River in 1541. The Mound Builders of the Mississippian, Hopewell, and Woodland cultures erected earthen mounds, some of which predated the Egyptian pyramids. Most of these people lived near rivers and made good use of their proximity to water for efficient mercantile transportation.

If the British navy reigned over the high seas, the French did their best to control the commerce of the inland waterways, relying, of course, on alliances with the indigenous populations who had preestablished trade routes for the Europeans to follow. Pierre Laclede and Auguste Chouteau "discovered" St. Louis, where I was born, and established it as an entrepot for the fur trade. The voyageurs and trappers (coureurs de bois) were a rugged lot, reaching their tentacles far up the Missouri River and into Canada, with the river as their main highway for trade. They paddled and portaged birch-bark canoes and ate pemmican out of their pockets—feisty sons of bitches, they drank and brawled with each other into oblivion at the yearly rendezvous. Their cargo of fur was a lucrative item for trade both within the colonies and back to Europe through the port of New Orleans.

To the east, colonists came across the Appalachians and found the Ohio River to be a more inviting highway than the existing roads. They built wooden "flatboats" for cargo and to take settlers progressively further into the west, finally uniting with the Mississippi River. Flatboats were a one-way means of transportation, not meant to be used for upbound commerce. When they reached their eventual destination, usually in Louisiana, the boats were dismantled, and the hardwood was sold as building material. Opposite of the Fats Domino lyrics, the crews walked back home *from* New Orleans or rode horses, often subjecting themselves to the depredations of outlaws who lurked on the paths to rob and murder them. Menacing as the Natchez Trace and other trails were, flatboat crewmen never fought against the current by attempting to go up the river.

Alternatively, keelboats were not dismantled upon arrival. They were coaxed back upstream through a variety of methods, none of which involved any kind of motorized assistance. Crews would use long poles that reached the bottom of the river as leverage to push the boat upstream. That was the preferred method, if it was shallow enough for the poles to find a bottom. Less popular was wrapping a line (rope) around a tree for mechanical advantage and pulling against it or, when all else failed, simply using the rope to pull the boat against the current from the shore. If they were extremely lucky, a following wind enabled them to use a sail. No wonder folk heroes like Mike Fink were embellished into legend subsequent to the evening ration of whiskey. Tough guys, but they didn't have long life expectancies.

If you were to have a successful river trip, no matter on what type of craft, you needed someone aboard who knew the river. Knowledge was mostly passed by word of mouth until Zadok Cramer consolidated the information into book form in 1801. Called simply *The Navigator*, it provided information on towns along the way and charts of the rivers, even the mercurial Lower Mississippi.

Inventors had been experimenting with steam engines since the late seventeenth century. In 1807, Robert Fulton successfully launched the *Clermont* in New York. The first functioning steamboat, it was soon to revolutionize water transportation in the United States. Here was a fledgling nation, with a mercantile and expansionist spirit, looking for an efficient vehicle to employ on a highway system provided by nature: the rivers. The steamboat was the perfect invention for the zeitgeist of the times. By 1811, Nicholas Roosevelt had taken the first steamboat, the *New Orleans*, down the Ohio and Mississippi Rivers to its namesake city.

Steamboats quickly became all the rage, transporting goods and people on the great and even the not-so-great rivers. Soon, new steamers were minted regularly from boatyards up and down the river. When settlers moved west, if they could afford it, they usually made at least part of the voyage via river. All manner of goods was shipped in steamboats and with greater speed and quantity than had been done previously.

One commodity in particular, cotton, became the linchpin of river traffic in the South. Cotton was a staple in the rapidly expanding nation and was also shipped to Europe from the Gulf of Mexico, where it was normally delivered by steamboat. Producing vast amounts of wealth, the shameful trinity of plantations, slavery, and steamboats became institutionalized over the face of the nation. Entrepreneurial steamboat owners made their money back on their investment within a few months' time, if the wooden boats they owned didn't sink or blow up in the process.

Further north, on the Upper Mississippi, acres of logs were floated down rivers, such as the St. Croix, the Chippewa, and the Black, in rafts. Steamboats were used to push and guide the loosely assembled and unruly rafts to sawmills where the lumber was turned into usable wood for construction in growing cities throughout the Midwest.

Hazards, such as snags, sawyers, sandbars, and rocks, were strewn along the steamboats' path. Added to that danger was the proximity of scalding boilers to wooden decks. And, of course, owners wanted to get rich before the boat sank, so shortcuts at the expense of safety were regularly employed to get the steamboat to its next landing as quickly as possible. Groundings, fires, explosions, collisions, and other calamities were commonplace.

Rules were few in the early days of westward expansion and fewer yet on the river. The steamboat captain was the king aboard his boat and ran the show as he saw fit, usually to maximize his profits. If the captain was the sheriff in this romantic trope, the pilot was the cowboy or gunslinger. Beholden to no man, his knowledge of the river was vital to getting the cargo and passengers where they needed to go. In *Life on the Mississippi*, Mark Twain presents an idyllic portrait of river lore and chic. Of course, he never writes about such practical niceties as toilet protocol, drinking water, or bathing. With little in the way of an internal plumbing system, one can imagine how those and other normal human functions took place. There was no Transportation Safety Board, Health Department, Environmental Protection Agency, or Clean Water Act to enforce orderly compliance. No government regulations—*laissez les bon temps rouler*!

On the Lower Mississippi, with a normal abundance of deep water, steamboats were built larger and grander by the year. On some of the tributaries, however, in low-water periods, captains were forced to lighten up on their draft to keep going. In the 1840s, principally on the upper reaches of the Ohio River, steamboats began to attach loaded barges alongside to lessen their draft without leaving valuable cargo behind. The first steamer of record to do this was the *Walter Forward* in 1845. The coal trade from Pittsburgh was the first recipient of this creative adjustment to low-water conditions.

Old rivermen's hyperbolic machismo created "barges of wood, and men of steel." At least, the first part was correct. The barges were made of wood and normally carried coal. Initially, the steamers towed the barges alongside. As most towboaters of today will scream, if given the opportunity, this is a noxious way to tow barges. They steer much better, don't bump into the boat, and improve air circulation when they are shoved ahead. Even the renowned river historian Captain Fred Way did not, from my research, pinpoint the exact time when the first boat with tow knees for pushing ahead was introduced on the scene, but it seems to have been an organic process that occurred over the decades.

As any contrarian will tell you, be wary when things seem too good to be true. Just when the future for steamboats looked limitless, two impediments reared their ugly heads. One was the advent of the railroad. Slow to join the party, since a railroad requires tracks while a river only needs water, trains could eventually go more places than the steamboat could. This was particularly true due to smaller rivers that would almost dry up later in the summer and fall or ice over in the winter. In some cases, rail and riverboat could complement one another, but normally, it was viewed as an antagonistic relationship. The second and more acute event, of course, was the Civil War. The river became a battleground that was dangerous for commercial navigation. One adaptation, however, that took commercial steamers decades to incorporate, did come out of the conflict. That was the metal used in ironclads, such as the *Monitor* and *Merrimack*.

At war's end, two significant events occurred that were emblematic of the steamboat era, one tragic while the other was steeped in glory and folklore.

The *Sultana* explosion unfortunately typified the steamboat owners' often cavalier attitude regarding safety amidst their gluttony for profits. In late April of 1865, the Civil War had recently ended, and Abraham Lincoln had been assassinated in the Ford Theater only days previously. Licensed to carry 376 passengers and eighty-five crew members, the *Sultana* left Vicksburg with an estimated 2,500 souls on board. Most of the passengers were emaciated Union prisoners of war returning home after the living hell of Cahaba and Andersonville Prisons.

Very few aboard realized that a Faustian bargain had been struck between Captain J. Cass Mason and the Union quartermaster Colonel Reuben Hatch in Vicksburg. Captain Mason and Chief Engineer Nathan Wintringer both knew that two of the *Sultana*'s four boilers were in need of repair, which, if done properly, would take several days to complete. Hence, only a superficial patch job was made so that the boat would not lose its human cargo to some of the other steamers in the vicinity.

A few hours north of Memphis, near Paddy's Old Hen and Chickens, the boilers exploded, and the *Sultana* became a torch of death. Many of the people were burned alive or scalded in the wooden cauldron the vessel had become. Others tried to swim to shore, but the river was miles wide and in flood stage.

Like a blue-coated flotilla, the bloated, burned, silent corpses eerily floated down past Memphis, some of the bearded faces frozen in expressions of agony and disbelief that they had made it through the war only to be unmercifully slaughtered on the steamboat taking them home. Seventeen to eighteen hundred people, mostly soldiers from the coal mines of West Virginia, the hills of southern Illinois, and the corn fields of Indiana, perished on April 27, 1865. Despite this wanton disregard for safety, no one faced any sort of criminal charges in the most deadly nautical disaster in US history. Yes, more people were lost that night than in 1912 on the *Titanic*.

Some of the grandest steamers were built immediately after the war. Captain John W. Cannon of Kentucky had the *Robert E. Lee* built in New Albany, Indiana, in 1866. It was a majestic steamboat in every sense of the word. Long as a football field with stacks seemingly spearing the clouds at 140 feet in height, the *Lee* had twin side paddlewheels thirty-eight feet in diameter. It had eight boilers and burned forty cords of wood a day, before it turned to coal as fuel. Even though slavery had been outlawed, cotton was still a valuable commodity, and the *Lee* could carry 5,000 bales of the white gold. Wealthy passengers enjoyed thick carpets, mahogany and teak woodwork, fine dining, and a spittoon every six feet out on the decks.

The *Lee* had reached such a level of fame that Captain Thomas P. Leathers became jealous. He had a boat constructed in Cincinnati and named it the *Natchez*, which was the sixth vessel of that name that he had built. His steamers paid homage to the Mississippi town, famous for the fact that its antebellum population had more millionaires living there than lived in New York City at that time. Soon, the *Natchez* and *Lee* were in stiff competition for passengers and cargo. The arrogance of these captains could only be satisfied by a duel or a race. In a duel, only one of them might have died. A race, with its inherent scofflaw attitude for safety, could lead to many fatalities. But such was the hubris of the day.

On June 30, 1870, a cannon was fired in New Orleans, and the two steamers raced into history—destination: St. Louis. The preliminaries to the event were similar to today's World Series or Super Bowl with wagers being made in barber shops, pool halls, saloons, maybe even outside churches. The race itself, though, was not much of a contest. The *Lee* took the lead right out of the blocks and finished five hours ahead of the *Natchez* in St. Louis on July 4. (That was certainly good planning.) What is impressive, though, is that the time of three days, eighteen hours, and fourteen minutes has never been bested by a boat of that size, 153 years later. Was the river shorter then? No, it was actually longer with more meandering bends. If one assumes, for sake of argument, that it was over 1,100 miles from the foot of Poydras Street to the foot of Market Street, and the *Lee* made it in a little over ninety hours, then the average speed was a bit over twelve miles per hour. Even the *M/V Mississippi*, with its 6,300 horsepower and a couple of light landing barges, could probably not beat that time. The current, however, was certainly much slower than today, with no dikes, revetments, or cutoffs to funnel the river's flow.

In a sad, but fitting, epilogue to this gallant story, neither vessel was on the river fifteen years after the race. The *Lee* burned, and the *Natchez* was retired and used as a wharf boat. Captain Leathers met an unusual fate for a riverman. He was run over by a bicycle on St. Charles Avenue in New Orleans and never recovered, passing a short time thereafter.

It was to be a dead cat bounce for steamboats, though. The railroads were more adaptable and able to reach more places with more ton-mile efficiency. It was also quicker and safer for passenger travel. As fewer and fewer people chose river travel, the towboat slowly began to phase out the pre–Civil War packet steamboat. In *Life on the Mississippi*, Mark Twain mentions observing the *Jos B. Williams*, in 1882, taking a large tow. Indeed, that steam towboat did take a tow over one thousand feet long and more than three hundred feet wide down the river in 1898. Like the grand *Sprague*, these boats were stern paddlewheel steamers. By the early decades of the twentieth century, most had switched over to square bows and pushed their tows ahead of them, as is done today.

The conversion to steel hulls from wood came slowly to the river. It was not until the turn of the twentieth century that steel began to replace wood as the primary building material of choice, for boats and for barges. At that time, the river industry had become depressed. There was a standing joke in St. Louis known as the "dog catcher line." This was a name given to unemployed river pilots who took jobs as dog catchers to support themselves until a pilot berth came along. Even though "Big Mama," the *Sprague*, was built in Dubuque, Iowa, in 1901, river traffic was at a nadir.

But transportation is a cyclical industry. President Theodore Roosevelt recognized the need for more efficient commercial infrastructure and spearheaded the Inland Waterways Commission in 1907. However, it was not until World War I that much attention was given to a part of the economy that was almost considered extinct. The predominance of the railroads had nearly put river traffic out of business, and the economy was suffering because railroads could not keep up with demand and were demanding monopolistic shipping rates as well. The federal government established the Federal Barge Lines on the Lower Mississippi and certain rivers in Alabama. Coincidentally, maybe, towboats and barges were being modernized to steel hulls, then to diesel engines. The first full diesel towboat, the *Harvey*, was built in Nashville in 1923. Strangely, diesel towboats were not subject to the same rules and regulations that steamboats had inherited since the days of Mark Twain.

With government subsidies in place, river traffic was rejuvenated. Locks and dams were needed on the Ohio and Upper Mississippi to maintain a channel, so those projects were completed during the 1920s and 1930s.

The *Jason* was the last sternwheel towboat built, in 1940, and by 1945, there was, essentially, an equal number of steam and diesel towboats on the rivers. The last working steam towboat, the *Lonestar*, finally doused its boiler in 1967. All towboats, then, were diesel, and expansion of the industry continued—actually, exploded.

With new 1,200-foot-long locks built on the Ohio and a stable nine-foot channel on the Upper Mississippi, the Illinois, the Ohio, Tennessee, Cumberland, and others, subsidized by federal tax dollars, the great river highway was grander than it was even in the days of steamboats.

So, 1972 is when an insignificant deckhand named Lee Hendrix went out to tighten rigging on a snowy Christmas Eve and became a puff in the whirlwind of river history. I put my muscle into every ratchet on the way out to the head of the tow. I paused out there in the quiet, looking at the few cars that remained on the River Road from Alton to Grafton.

I pulled a Lucky Strike from my shirt pocket and struck a match to it, then sat on a timberhead, satisfied with what I had done. Soon, other small pinnacles of light could be seen approaching between the strings of barges out toward me. Then a voice could be heard.

"Whachu doin' out here, boy?"

It was the new mate, Larry, and the other deckhand. I worried I was in trouble for taking a break.

"Jus' tightenin' tow," I responded.

"Well, ya done a purty good job of bonin' her down. But come on back ta the boat, son. It's cold out here. 'Sides, I can't have no green deckhand fallin' in. Your mama'd wanna know why we didn't keep better keer of ya. An' it's too much paperwork for the captain."

I threw my cigarette butt into the snow and headed back to the boat with them.

"Son," the mate said, "ya might just make a riverman someday. Ya 'pear like ya kin handle the lonesomeness of it."

# CHAPTER 7

# Lock on the Head

There are twenty-six locks on the Upper Mississippi River between St. Paul and St. Louis. If you don't learn to embrace making them in rain, thunder, sun, snow, and heat, you should hop on the first chicken bone back home and quit posing as a deckhand. When the hard times come, and they will, seeing yourself as part of a winning team helps to pull you through. In 1975, I considered myself blessed, at least while I was on watch with Freddie and Smokey.

Freddie had lived most of his life on water. Born on a bayou near Lake Charles, Louisiana, he mixed just enough French into the jambalaya of his vocabulary to confuse us, which he did merrily and often.

Though Smokey seemed to pursue no particular destination in life, he was driven to charge full on until he arrived there. I was one of the few people on the river who knew that Matt Buttrick was Smokey's legal name, and that was only because I was the mate on the *Julia K.* and saw his name written on payroll reports. He was like many other deckhands on towboats in the 1970s: no one had known him before, where he had been, or what he had done. As in the French Foreign Legion, it was considered inappropriate to ask too many questions regarding the past of others. In his case, however, you couldn't help being intrigued. It was like the fascination one has at seeing the northern lights on a clear fall night near Lynxville. You stand and wonder about what unique atmospheric conditions have caused their sudden appearance and how long they will last.

Why Smokey lived and died the way he did was, and remains, his own business. There are hundreds of events that bring his image to my mind, but the most vivid was an experience we had at Lock 10 in Guttenberg, Iowa. It was to be the last lock he, Freddie, and I would ever make together. I can still, almost fifty years later, remember the God-awful fragrance of Grafton Johnnie's cologne as I came down the stairs to announce that the lock was "dead on the head." Johnnie was perplexed, having a difficult decision to make.

"I don't wanna miss ol' Sanford an' Chico tonight, Lee. Ya think there's gonna be any women there at that lock?"

"Why, hell yeah, *mon ami*! Dere gonna be *beaucoup de femmes*!" Freddie's voice bellowed from down the hall. "You don' wanna be waitin' in here smellin' like no New Orleans whore watchin' Redd Foxx. Brought yourself out dere wid us, man."

"Welll . . . Maybe I'll just wait 'til you get on the wall."

"Son, you don' wanna did dat. You stayed in here, we gonna make it too fast, *mon ami*. You didn' wanna disappoint dem snow-digger gals what come from all 'roun' I-oway to smell you. 'Sides, me an' Smokey, we need your company."

Johnnie fell dutifully in behind us as we walked into the deck locker where we grabbed life vests and headed onto the tow. Late summer on the Upper had brought its subtle but definable changes. The days no longer felt as though they would last forever, so we tied long-sleeved shirts around our waists for later in the evening. The deckhands' main plague—the willow fly—was gone for good or, at least, until the following June.

This tow was our work of art, Smokey, Freddie, and I having laid just about every wire out there ourselves. Tons of cold, hard rigging held together the even greater weight of 25,000 tons of corn, wheat, oats, and barge steel. Every fore and aft, jockey, backing, and towing wire had some of our sweat, blood, and choice vocabulary mixed in with it. We took pride in our work, especially when we could hear those fifty-four sets of ratchets, wires, and straps sing and pop when Jeremy or Jodie, our captain and pilot, steered around Coon Slough or Lansing. It was living proof that we had done something right.

Smokey had his usual "lock equipment" with him as his long legs nimbly bound over cavels, chocks, timberheads, or anything else that presented an obstacle to him. Actually, there were no obstacles for Smokey—everything in life was merely something to be dealt with and gotten through. Most of his lock equipment was safely stashed inside his "lock box"—an old US Army ammo box topped with a hand-cut foam cushion for sitting on. The box was decorated with a Grateful Dead sticker, then painted with orange, pink, and Day-Glo green designs. Inside the box were his guitar pick he had carved from driftwood, his sunglasses, head-strapped flashlight, and his latest reading material, which could range from Kurt Vonnegut to Hunter Thompson to *Zen and the Art of Motorcycle Maintenance*. Around his neck was the strap that carried his five-stringed Eko guitar he had purchased for ten dollars at the second-hand store in Winona.

Some people had "suggested" that I search his lock box for contraband because they couldn't believe that Smokey was naturally the way he was. I never did this, though, because I was confident that he was straight—no drug could produce a treasure like him.

Johnnie was showered and shaved with clean blue jeans and a purple tank top over his lean body. He looked like he ought to be back in Grafton sitting on the hood of a pale blue Malibu convertible drinking a Budweiser whose five buddies were kept reasonably cool on the floorboard out of sight of the local cop. "Smokey, you gonna try to talk to some girls while you're on the first cut?" he asked eagerly.

"I'm afraid I lack your propension for narcissism, John."

"I wisht for once you'd jist talk plain ol' English."

"My tan line's in the wrong place."

Freddie had already checked the lock lines at the break coupling when we arrived. "Dose cracker jacks on dat after watch done already moved dem lines over. Man, dey not backin' up to dere money dis trip. *N'est-ce pas, jolie fils?*" He laughed and ran his hands through Johnnie's carefully plumed hair.

"Hey," Johnnie said, ducking away, "Look, but don't touch. Told you, Lee, that after watch is on the ball. We know Lock 10's on the other side of the river from 9. I'm gonna watch you clowns and see if you can make a lock as good as we can."

"You boys better kep' it up," said Freddie, "or dat mate gonna gave you some o' dat Cajun jujitsu I already been teachin' him."

"Now, what the fuck is Cajun jujitsu?"

"That's a euphemism for a size nine up your rectal cavity." Smokey chose to enlighten him.

"A-you-fo' what? Man, talk English. Freddie talks it better than you do, an' he grew up wrastlin' gators on a bayou!"

We reached the head of the tow almost on the run as Freddie went immediately to our 240-foot lock line and began laying it out flat on the deck. Some red-winged black birds flew purposefully across the cargo boxes in search of an evening dinner of loose grain, which always managed to slip out onto the decks. Smokey moved our green starboard navigation light out of the way, knowing there would soon be some fast action there. I brought the speaker over from the center barge, got up on the cargo box, and waved it back and forth over my head.

"Hey," I said into it. No answer.

"Hear me, Jeremy?" I tried again. Still no reply.

"I wish to hell he'd turn that damn thing on. He's probably bull shittin' with those lock men or something."

"He like to did dat," said Freddie as he danced to the other side of the tow with a bumper. I was getting aggravated.

I held the speaker in front of my face and yelled at the top of my lungs, "*Wake up back there!*"

"What are you trying to do, break my ear drums or something?" Jeremy's voice crackled from inside the speaker. Finally!

"Hell, I thought you were asleep back there," I complained.

"Who could sleep with all that racket going on? If you guys got paid for how much noise you make, you'd all be millionaires and not working on towboats."

"Hey, Cap," said Johnnie, grinning like a possum, "if they got paid for how much work they do, they'd starve to death."

"Who is that out there? A spy from the after watch? You better be careful or you're going to miss *Sanford and Son*."

"Already missed it, Cap. Out here to see some girls."

"Well, you just watch yourself. Remember, you're representing the crew and the company."

"That Jeremy's somethin' else!" Johnnie again wistfully displayed his white teeth of youth.

"He's serious," I said.

From behind us, an abrupt exhaust blast announced the appearance of a long, sleek fiberglass speedboat, the type that makes so much noise it scares away every blue heron within ten miles and whose wake sinks fishing boats and threatens the stability of large islands. The driver of the boat was a sleepy-eyed chap with shoulders and a back that were so red they would bring tears after midnight when the booze wore off. His ample girth hung merrily over his shorts as he maneuvered the boat alongside our tow and to within fifteen feet of us.

It was his companion, however, who drew more attention than he. She had long, blonde hair, and her pink bikini was made even brighter by the golden color of her skin. She sat next to the driver with one leg slung drowsily over the side of the boat. She smiled at us as one does when studying exotic species in a zoo. Our murky existence would provide her with something to talk about at work come Monday morning.

Mr. Thundergut turned his engine to an idle, reached behind him, and pulled up a six-pack of Pabst Blue Ribbon that he held out in front of us. His first attempt to stand was unsuccessful, but he finally put one hand on the lady's shoulder and jacked himself up. "Hey, you guys want a beer?" he belched.

Oh, the weight of temptation! Had it been dark and the tow speaker not on, who knows how much strength we could have mustered to resist. We didn't say a word, though we all looked their way—these two were a hazard to safe navigation. Johnnie was, of course, the most interested. He started to open his mouth, but a quick elbow to the ribs silenced him.

"Don't even think of it," I told him.

Thundergut was not too easily put off, however. "How 'bout a blonde?" he yelled. She laughed uncomfortably, having doubtlessly read some sailor books in school, and plunked Thundergut unceremoniously back into his seat.

He revved his engine so that it could create hearing loss in Cassville and sped away.

"*So long, chumps! Drink beer, get naked, rock and roll,*" his voice somehow came over the din.

"He expresses his hedonistic desires quite succinctly," said Smokey.

"Au revoir, Jolie Blonde."

We all glared after them with some degree of envy but soon realized that, with his impending sunburn and hangover, not to mention the escalating price of gasoline, we'd all feel much better on Monday morning than Thundergut would. Besides, we had other things on our plate, such as Lock 10. Late summer rains had swollen the river, and the gates of the dam were "on the hooks." It was open river, creating wicked crosscurrents and eddies all around the lock.

Our tow was one thousand feet long, 105 feet wide, and weighed probably 25,000 tons. The customary way to get the tow into this lock when the river is as "hot" as it was is to come down so close to the concrete guide wall on the Iowa bank that you miss it by less than five feet. You can't slow down too quickly because, if you do, the whole tow will be taken by the current and head out toward the dam like a scalded dog.

That's where the skill and concentration of the captain and the deck crew come in. Jeremy would not be able to see all the way out to the starboard corner of the tow, so I would be his eyes. I would communicate distances over the speaker to him, at first, every thirty seconds or so, but when it got down to show time, my voice and his hands would have to act almost in unison, and I'd be talkin' like there was no tomorrow. He had to trust my judgment, and I had to trust him because I'd be the first one on the firing line to get hit if something went wrong.

"Twelve hundred feet above and thirty feet to the bad," I said into the speaker. I was standing down from the starboard corner of the tow with my right foot about three inches from the side.

"Okay, Mr. Mate. We'll have to hold her in pretty far behind the wall."

In other words, if we kept going straight, we would face our 25,000 tons dead into the concrete lock wall.

The liquid sounds of the tow shoving through the water became slower and more pronounced as our headway decreased. The sloshing became rhythmical, and the crickets and tree frogs added harmony to a late-summer evening concert.

"A thousand feet above, still holding thirty to the bad."

"All right, mate. Be careful—you've got more spies from the after watch coming out."

"Who's that?"

"Oh, it looks like a pilot and an engineer. They seem to be in a hurry, running like rats, you might say."

"Hey, Cap," yelled Freddie. "Dey goin' up to Moxies for some o' dat hoochapappy [i.e., booze]!"

"Well, you tell them not to get too much of that hoochapappy!"

"Four hundred above, twenty feet to the bad."

"Got you fine, Mr. Mate." We rested with our elbows leaning on the barges: this was the calm before all hell broke loose. Twenty seconds passed.

"Two hundred, still fifteen to the bad."

"I have that alright."

"One hundred now. Still ten to the bad."

"She'll start coming out soon, son." We were getting primed.

"Fifty above, five to the bad!"

"Alright."

"Twenty-five, three to the bad!"

"She's coming out now. Bless her little heart."

Freddie hung a bumper, about four feet long and over twenty pounds, over the corner of the tow, just in case the current would forget to do what it had done since the 1930s when the lock was built. Two burly lock men stood at the end of the wall with their small handy lines swinging at their sides as an invisible broom of water swept us out around the end of the concrete wall when we got within fifteen feet of it.

"Just breakin' daylight!" I yelled as we missed the end of the wall by a good three feet. Then the fun really started.

Both lock men took the monkey fists on the ends of their small handy lines and hurled them down to Smokey, almost crowning him. He laughed and took the first one and draped it through the eye of our thick, long lock line. As we continued to float alongside the concrete guide wall, we fell further and further from it. The unluckier of the two lock men, whose handy line had been grabbed by Smokey, hauled our line down the wall and began puffing and chugging with it as Smokey fed him slack.

We had another three hundred feet to go before we'd have to squeeze our 105-foot-wide tow into the 110-foot hole. The head of the tow, by this time, was about thirty feet wide—and steadily falling out. The short, yellow-painted "bullnose" hung menacingly ahead of us. This was the part of the show where an Upper Miss deck crew earned its cornbread.

"Any time you're ready, Cap!" I yelled.

"Let's get the next pin, mate."

The lock man thankfully slung the eye of the line on a solid yellow mooring pin anchored to the top of the wall; then, he and his partner, with their remaining lung capacity, got as far away from the line as they could get, put their hands on the nearest guardrail, and tried to catch their breath.

Smokey took the bight of the line and wrapped two entire crisscrosses around the deck cavel at his feet. Freddie backed him up, making sure Smokey did not get a foot or leg caught up in the loop of the line. The line quickly became tight and began to menacingly crack, ping, and chortle around the cavel. Smokey did not attempt to hold the line too tight for it would have broken. His artist's feel allowed him to reach the subtle point where he was just checking the swing of the 25,000 tons as Jeremy backed the boat for all she was worth.

"Don't break your line. Let me stop it." Jeremy's voice was soft and nurturing, like your mother's was the first time you cried in your crib. Smokey's arms began to bulge, and he breathed heavier. By this time, we were only 150 feet above the bullnose and a long way from being able to fit down in that hole. But the swing of the tow was becoming slower.

"Alright, son. I've got it stopped. See if you can hold the head."

Smokey lustily threw two or three more wraps on the cavel and backed completely away from it. Our line, made of synthetic nylon and other assorted plastics, tightened as the entire weight of the tow came down on it. The fibers creaked and fused themselves together. The line, which had started in a two-inch limp diameter, was now about as big around as a cane and so tight it looked like you could have cut it with a butter knife.

The line was our friend and ally, but it was an uneasy truce. If it should happen to break under this pressure, it could cut a swath, like a hurricane, in its deadly wake. A broken poly-d lock line can lop off limbs of deckhands like lumberjacks do on old trees. But at the moment, she held tight as we stared right down at the open gates of the dam with all the power of the Mississippi trying to send us that way. Hard to believe, but the situation was under control. For the time.

Jeremy slowly started to work the stern of the tow toward the guide wall. Once he got it over there, we would not have the current between the tow and the wall but on our port side, and we would begin to come back toward the guide wall.

Freddie had regained his breath and joie de vivre. "Hey, Smokey," he said, "what you tink, man? If I gave you five dollar, you crawl over dat line to de lock wall?"

Smokey mused for a few seconds. "The line has reached its maximal torque and tensile-strength utilization, so I would gladly accept that wager, were it not in direct violation of company safety standards."

"Oh, no. Don't tempt him. He might do it!" Jeremy spoke only half in jest from the speaker.

The principles of leverage and motion had begun to take over. Smokey could probably explain the physics of it, but all I know is that when the stern comes in and the line gets tight leading down the river, the entire tow starts to go toward the wall. The trick is to time it just right so that you let loose the line and shove down into the lock before that massive weight hits the wall, bounces out, and just generally spoils your plans.

The head swung in slowly at first. Then, like an old hog sensing dinner, picked up speed, and concurrently, we eased down toward the bullnose. Strange how relationships change, but that 25,000 tons that had worked so hard against us was now going our way. We were forty feet wide and less than one hundred feet above the nose; then twenty-five feet and eighty above; then ten feet wide and . . . "Turn her loose, son."

Smokey quickly unwrapped the line, Freddie grabbed a bumper, the lock men sprang into action taking the line off the pin. We slowly picked up headway—the tow swung in and gently touched the wall. The wheels of the boat spewed up wash as our momentum enabled us to outrun the draft toward the dam, and into the lock chamber we went.

I took the time to check my watch. We had been on the head of the tow for almost a half hour. *Sanford and Son* was over.

Once clear of the gates and inside the lock chamber, everything was peaceful, and the water was slack and calm. My voice, trapped within the confines of the 600-by-110-foot chamber, echoed off the concrete walls and back at me. Freddie and Smokey dropped their bumpers and headed back to the break coupling while Johnnie looked for a ladder to hop off on. Jodie and Hairy were already sprinting up River Street toward Moxies. They knew they'd have to hurry because it would be a pretty quick lockage with the dam wide open.

We came close to a ladder, recessed in the wall, and Johnnie hopped off. He was in luck because the viewing gallery was packed, including some teenage girls with whom he would likely strike a conversation. The people in the gallery stared at me as I talked into the speaker. I was self-conscious, not knowing if I felt more in a parade or a zoo.

Our tow was too long to fit in the chamber, so we had to make what is known as a "double lock." Our "first cut" of nine barges, 595 feet long, would be separated from the remaining six and the boat, then locked through and pulled out with a cable attached to a winch. Jeremy stopped the head of the tow a few feet from the lower gate recess as we eased down into our first line. I hurriedly put my backing line on, fighting the side coaming of the barge, then took my gloves off and prepared to catch a bit of a breather.

The lock man, a balding, fiftyish Guttenberger know to us as Buttons, leaned over the wall to talk. Back at the break coupling could be heard the racket of chains and ratchets flying across the decks as Smokey and Freddie kicked them off. Were it a foggy morning about 3 a.m., those sounds would be heard on every street in town, which tells you a great deal about how sound carries on a foggy river morning—or tells you something about the size of Guttenberg, Iowa.

After a few minutes, the racket abruptly stopped, and Jeremy backed the second cut out of the lock. They tied it on the upper guide wall. The upper gates were closed, and our nine little darlings were in the lock by themselves.

"See yuse still got dat crazy Smokey on," said Buttons, who was gratefully cozying up to a government pension.

"Wouldn't know what to do without him."

"Ever check any of his care packages for wacky tabacky?"

"Careful, now, we'll put him off and turn him loose on Guttenberg. No, believe it or not, he doesn't even drink beer."

"Oh, yaaah. How'd he git so damn weird den, fer Chrissake?"

"He's a hell of a deckhand. He fears nothing except questions about where he came from."

"That's what's so goddamn hard to figure out." He shook his head. "I was talkin' to a buddy o' mine works up at Lock 9, said that crazy son of a gun come out wid a burlap sack 'round him an' a pair o' dem Jesus boots ta make a lock once. An' I says, 'Geez, I don't doubt it, knowin' him!' Looks like he reads a lot o' books. He been to college or somethin'?"

"One of the questions I wouldn't know how to answer. I do know the kids down there'll love to hear him play his guitar."

"How'd he come by that nickname? Fireman or somethin'?"

"He likes to get close to flames and climb out on bridge girders. I do know that."

"Jesus fuckin' Christ! One o' dem, huh? Takes all kinds, don't it?"

Buttons would periodically check his panel board and spit in the river. Finally, he said, "Okay, yuse guys are 'bout down. We'll get ready ta pull yuse out. Be careful down there—had somebody lose a first cut last week."

"Thanks for the warning." It was time for more of the fun and games that make towboating exciting. They would pull a long cable out, drape the eye over the wall, and attach it to a deck fitting to pull out our first cut with an electric winch. That part was easy enough, but the action would start when the current grabbed those nine loaded barges, and we had to stop them with only our lock lines for brakes. Smokey and I would be on the barges, so it was in our best interest to get them whoaed down or we would get a free, all-expense paid trip down the Mississippi without engines, rudders, or meals until we ran headlong into a dike or sandbar. I reckoned that when I became a pilot, making big money, putting on weight, and getting soft, there would be times when I would miss this bit of head clearing, living-for-the-moment physical challenge. I hadn't reached that point yet—I dreaded it! On the other end, "riding shotgun," was Smokey, and he thirsted for it.

The lower gates cranked slowly open as Freddie came running down the wall. It was his job to help the lock men pull their cable out. They always got pissed off if they had to pull it out by themselves, like they would get a hernia or something.

"Hey, Mac," said Freddie to the younger of the two, "want me to get you some more o' dat Cajun coffee?"

"Jesus Christ, no! I made a pot o' dat stuff an' I couldn't sleep for three days. You coulda floated a horseshoe in dat!" he returned, puffing painfully with cable in hand.

"Man, you snow diggers don' know what good coffee is! Gimme dat cable, Mac, I'll did dat." Mac gladly stepped aside. "Here, Lee," Freddie said, handing the eye down to me.

"Your buddy ready back there?" Mac asked, looking uneasily toward Smokey's end of the cut six hundred feet away.

"Dat *ami* o' mine, Smokey, he stay ready!"

"He don't need me ta run back an' throw his line off the wall, does he?"

"Well, now dat, I can' say. He might did you like he did ol' boy up at Lock 5. Dat boy, he wouldn't go back ta git Smokey's line, so Smokey, he lef' it on. Dey goes to pull dat firs' cut out an' dat cable was jist a-spinnin' an' a-groanin' an' dey can' figure out why dem barges not comin' out. Ol' Smokey was jist a-hootin' like a barn owl back dere. Figure dat lock man not doin' his job."

"Jesus fuckin' Christ, I guess I better go back there then," Mac said, like somebody who just found out he had to work overtime.

"Crazy sumvabitch," spat Buttons as Freddie winked at me with a "dese-Yankee's-kinda-silly" grin.

Just then, the horn blew. Freddie took my line off the wall, and the engine on the winch began to pull the cable taut. Slowly as it ground, the barges began to ease out of the lock.

"Phhhhhht!" Mac whistled worriedly back at Smokey. Buttons annoyedly threw his hands over his head three times. Smokey's unpredictability kept everyone on their toes.

I looked back as Smokey gracefully rolled his eye off the top of the wall—something not everyone can do. Mac and Buttons were greatly relieved.

As the barges cleared the lower gates, I handed the eye of my lock line up to Freddie. The sun was gone for good, not even a last splash of pink and orange off the trees and water. The world appeared in a black and white pallor. From down the river around Goetz Island or St. Louis Woodyard, I could smell the evening's first campfires.

"Don' forget dem love letters!" Freddie said, pointing to the three or four bent, twisted envelopes in my hip pocket. He walked on the lock wall keeping pace with me as the barges floated down.

"Thanks for reminding me."

"I figured one o' dem's for Peggy. Hell, you might be married nex' time you come back."

"Shoot, not yet!" I said, sometimes wishing it were true. "That's what you need to do: settle down."

"Nooo! Cain' did dat, me. River life ain' no life fer no married man. Make tings too complicated. Thirty days, forty days—dat's long time man's gotta remember all dem promises he made."

"'Sides, you hafta spread yourself around," I teased.

"Hey, las' time I'm off, dis lady fren' o' mine in Lafayette, smart woman, her. She show me dis poem. Fella say de river is a strong, brown god. What you tink 'bout dat?"

"Don't know—'cause it's powerful, maybe? Sounds like you bin thinkin' on it." I wasn't into literary analysis at that particular moment.

"Me, I tink maybe could be 'cause de river, it bring life an' don' ever stop. 'Cause when you give your life to de river, it'll always take you back—so you always go back to it. You tink you could see de river, but de part you cain' see—why dat's de strongest part dey is, de part dat grab hol' an' won' let go."

Interesting though this discussion was, the time for philosophizing had passed. The barges had reached the sign on the wall that read "900," and their flow had picked up. Freddie was now at a steady walk to keep up with me. "Let's give 'em a check," I said.

He threw the eye of the line on a pin, and I began putting wraps on the cavel. As soon as the line got tight, I held what pressure I could. The line began to talk to me—through my ears where I heard the pop and in my forearms as I fed out the slack. I checked for fifteen to twenty seconds, running out over fifty feet of line. The barges had not slowed down at all.

I began to get worried. Maybe I should have started checking earlier. Well, it was too late for self-incrimination. The one thing I did not want to do was break my line, so I threw my wraps off and told Freddie to grab the eye. He ran the eye down the wall toward me as I pulled in the slack with my forearms, which were starting to feel pumped. Then the line dropped in the water, which made it heavier to pull up, and my arms started to lose feeling. From the stern, I could hear the popping of Smokey's line as he was checking, too. We still weren't slowing them down, though. Soon, Smokey would be by the gates with no option for checking until he cleared them.

Maybe if we both checked together, I thought. "Let's double it up," I said to Freddie. Time was of the essence because Freddie was near the end of the long wall, and the pins were fifty feet apart. He handed the eye back to me and put the bight on the closest pin. The strain of two double lines on the barges began to slowly diminish their headway, but I was holding my line much harder than I wanted to. Then it happened.

The effects of a wet line with too many wraps around a small cavel created enough seizing and fouling to bind the line down on itself. I jerked three or four times on it, but it was hopelessly fouled. The only thing to do was to retreat—get away from it and get my other line ready. Adrenaline was working overtime, and I held onto the line a little longer than I should have. The popping and crackling sounds slowly changed to twisting and tearing. Sweat cascaded from my face and arms, though my mouth was strangely parched, as I screamed, "*Run, Freddie!*" The energy within the line exploded like a shotgun blast. Floating line fibers filled the gloaming night air as the line lay broken and limp at my feet. My arms had long since lost any sensation, so I had to visibly check my appendages to make sure everything was still there. I could look up and see that Freddie, thanks to God, was still alive and on the lock wall.

I grabbed the other lock line I kept on the head and threw the eye up toward Freddie. The tip end of the barges was just passing the end of the wall. We were down to our last pin. The barges had started to slow down, and the sounds of Smokey's line six hundred feet away let me know he had cleared the gates and was again giving it all he could. If I could just give him a little help, we could still get them stopped. Freddie put the line on the pin, but the parts of my broken line had fused and melted themselves together on my cavel. The only way to get them off would be with a fire axe, and I had no time for that, so my main deck fitting was useless to me.

What I could do, though, was drag my line down the side of the barges to the next cavel. Of course, this wouldn't be too sweet either because I'd be going from a working area that was as big and wide as the Gobi Desert to one like the inside of a cave, and

I'd have to drag that anaconda-like line down the side and be careful not to get drug in the river.

Smokey had them slowed down to a crawl, but he had to be running out of line. The power of the river was deceiving—the last little bit was the hardest, the part you can't see—the great, brown god. I finally got my line in position and readied myself for battle. There was some old, mashed up corn and oats that made the decks slippery and was lumped around the cavel that I had to use—the price you pay for towing grain. Added to this messy inconvenience was the fact that the cavel was small, and my line would bind quite easily. But I was in no position to be overly choosy.

The lead on the line would go right by the barge coaming, so that if I stood between the barge and the line, I would be pinned in between them. I had to position myself downriver of the line, which meant that if this one broke, I would be standing right in its path. I cursed the life jacket that I had on, making it much clumsier for me to work in such a confined area. I threw some wraps on the cavel, held tight, then hoped and prayed for the best. If this didn't work, we were going down the river. Of that, there was little doubt.

My line tried to foul. I whipped it around as best I could, but it didn't do any good. It was stuck fast. However, the crackling of Smokey's line was getting less and less pronounced, and the barges were just barely moving. My line would either hold and stop us—or break. I hunted for an avenue of escape if it came down to raw survival as the barges took one last lunge—then stopped. The line held. We had done it!

From the stern of the cut, Smokey let out a long "*Ooooooh-aaa-hahaha!*" at the top of his lungs, likely causing impromptu reaction from the viewing gallery. They were getting their money's worth from us. Smokey was in heaven, tripping on adrenaline. I was somewhere quite different because I still had to figure out how to put a backing line on so the barges couldn't surge backwards. My arms felt like they were connected to two ship anchors.

I hurried to the head, heaved what remained of the broken line on my shoulder, and huffed two hundred feet back to the first barge coupling. Freddie helped me to get a backing line on, and I searched for a ladder to climb up the lock wall on. My arms ached so badly, I could barely climb up the ladder, and though the night had become much cooler, my face, hair, and upper body were soaked with sweat.

The incandescent lights atop the poles high above the lock walls were burning as Freddie and I made our way back to the second cut, a distance of about 1,200 feet. Through the thickening night air, we could make out Smokey's guitar, then his voice:

> If you're looking to get saved,
> You better go back to from where you came
> 'Cause the cops don't need you,
> And, man, they expect the same.

I grinned down at him, sitting on his lock box, as I passed, and he winked back without stopping his song. He would sing and play until we were ready to make our coupling again. How his hands had the strength to grip even a guitar pick or how he had breath to sing was a mystery to me.

"Didn't think yuse was gonna get 'em stopped," said Buttons as we strolled by.

"Aww, man, had 'em all de way," said Freddie. "I bet *mon ami* Smokey had his line a-smokin'."

"You betcha, crazy sumvabitch. Listen at 'im now."

"*Oui, monsieur. Douce musique.*"

I was thirsty, so I reached into my pocket and grabbed enough change to buy us all a coke at the friendly machine all the locks had. Grafton Johnnie was busy telling a teenage farm girl from Dyersville that he was a deckhand but was preparing for "pilot training." She appeared suitably impressed.

I would have to walk by the viewing gallery on my way back to the second cut and put myself on display to everyone again. There was nothing easy about decking on the Upper and not much of a way to stay clean through it all. Your White Mule leather gloves, cutoff blue jeans, sleeveless work shirts, and headbands were blackened by grease from ratchets and wires. You carried odors around with you, too. Grease, sweat, suntan lotion, sour corn mash, poly-d lines, and a little torn flesh were a deckhand's potpourri. Not pretty, but proud smells. And when a soft city boy, like myself, felt biceps, forearms, and chest muscles gradually start to harden, you became more likely to strip the sleeves off your shirt and show your scrapes, cuts, and bruises to anyone that wanted to look at you. And when people at lock walls glanced sideways at you with the sort of wonder that one displays toward that enigmatic personage who has been wild places and done wild things, you thrust your nose out in pride. Who cares if they look on you as an outcast or a criminal—the more they wonder about you, the better you like it. Hell, they don't know that you read Tolkien and listen to Beethoven in your spare time.

A farmer from Allamakee County, down to Guttenberg for the weekend, leaned his head over the railing. "Whatcha got in dem barges?" he asked me.

Just for a moment, I wished I had the audacity of Shorty Kelly. Shorty would strut like a banty rooster, chains from his wallet in his back pocket jangling, arms displaying tattoos of snakes curled around bloody knives and skeletons. When asked this question, he would stop, put his Marlboro reflectively between stubs of brown teeth, and declare, "Fifteen loads of Vaseline for the Virgin Islands, partner!" and prance authoritatively away.

"Dumb ol' bastard!" he would laugh as his body bounced bodaciously along on his duck-toed legs. In mid-stride, he would lift a leg up and, *fffffttt*. "Not bad manners, just good meatloaf," he would laugh. Well, on second thought, Shorty wasn't such a good role model.

"Corn and beans," I replied to the farmer.

As Freddie and I jumped on the second cut, Jeremy's voice raked out of the tow speaker. "You had me worried for a while. I thought

I was going to have to chase you down the river. Bet you couldn't have drove a six-penny nail up your ass with a sledgehammer when that line broke."

"Wouldn't dream of spoiling your night like that."

It was almost completely dark with just a faint touch of orange in the west. Layers of blue and black clouds draped over our heads. I leaned against the steel of the barge coaming and did what a riverman does best for a few seconds until the blast of the lock horn broke my trance. It was time to go in with the second cut.

The viewing gallery had quickly emptied with the curtain of dusk, people having come up with better options for the night's entertainment. Even with the waning of summer, dusk also brought mosquitoes. The girl from Dyersville slipped Grafton Johnnie her phone number. From the lower end, a shouting match was going on. I couldn't hear the details, but from past experience, I knew that it was between parents anxious to leave and their children, who knew nothing about Bob Dylan but were happy listening to Smokey herald in the night.

The second cut was quickly and uneventfully locked through, the gates opened, and the horn blasted. We eased the second cut down to couple our tow back together. It was further down the wall than normal due to our little adventure. Smokey and Freddie firmly fed the seven-eighths-inch wire cable around chocks, timberheads, and cavels into the seventy-pound ratchet. There, in the stillness of the late summer night, on a stage of corn and beans in northeast Iowa, they were ballet dancers in a whirling, spinning pas de deux. They grabbed the ratchet, jerked, and hooked just the right chain link. From here, it was a gravy train.

"Port wire, she hooked," announced Freddie.

From somewhere across the lock wall, out of sight, we were serenaded by Captain & Tennille singing "Love Will Keep Us Together" as we now worked silently, each man familiar with his role. Sometimes we worked briefly in pairs, often alone, but still a team, a family. We sweated, grunted, squeezed, pulled, and jumped on wires, straps, and ratchets. We put ten-pound toothpicks in the ratchets and tightened them with long "cheater pipes" so not an ounce of slack would remain in that coupling.

After about ten minutes, I took one final lunge with a cheater pipe, looked around, and yelled, "We're ready!"

As if on cue, Hairy, Jody, and Grafton Johnnie scurried down the ladder, and this lock was fixin' to be history.

"Well, turn me loose, Mother Goose." Jeremy blew the boat whistle and backed up so the taut lines could be released. Mac threw off our head line. Buttons took our stern line off, walked it back to the coupling, then produced a camera from his pocket.

"Don't know if this flash'll work, but my granddaughter wants me to get a picture of some of yuse deck crews so's she kin show ev'rybody at school what her grampa does." Ansel Adams he wasn't as he fumbled with the camera. We molded together, Freddie in the middle, one arm around my neck and the other elbow on Smokey's shoulder.

"Hey, Buttons, you tol' dat little gal dis de bes' deck crew what dis river got. An' dis guy here, Smokey, why he de bes' *ami* from what I ever had. He's a hero."

"Hero?" laughed Buttons. "Yuse take care o' dat Smokey, now," he said as the light flashed in our eyes.

"Well, if you movie stars are finished posing, how about turning my running lights on." Jeremy's voice came out of the speaker.

With the appearance of a red, green, and flashing yellow light as our coda, down the river we went. The next day, my relief showed up at Bellevue.

A few weeks later, with my time growing short before I would be called back to the *Julia K.*, I was drinking coffee and reading the local newspaper when I was drawn to a brief, fifteenth-page filler article among the tire ads.

DRIFTER PERISHES IN FIRE AFTER RESCUING CHILDREN

A man registered in a transient hotel under what is thought to
be an assumed name is credited with miraculously rescuing several
children from a raging inferno at a residential children's home on
Chicago's south side.

The fire, which started in the early morning hours quickly
engulfed the building, making inside evacuation impossible.
According to the children, who ranged in age from four to ten, the
rescuer appeared through a window and was able to lower them
with a rope, one or two at a time, to the ground.

The man had somehow traversed the short distance from the
Eutopia Hotel, where he had been staying for two weeks. After successfully evacuating all of the children living on the upper floors,
the man perished while trying to traverse back, rather than come
down the escape route he had made for the children. Firemen who
eventually arrived on the scene could give no reason for his fatal
actions. They stated that, if it had not been for his heroism, most, if
not all, of the children would have likely died in the blaze.

An employee of the Eutopia stated that the man, registered
under the name of "Billy Pilgrim," had spent a good deal of his
time making friends with the children, playing guitar, and singing
with them. Other than that, no information about him is known at
this time.

Smokey wouldn't have had it any other way.

When I returned to the *Julia K.*, we were making our last trip
out of the Upper before the ice season closed in on us. The ash,
maple, and oak had shed their leaves, and days were short, framed
in castles of gray above our heads. We coiled our lock lines around
oil lamps or stuffed them in barge holds to keep them from becoming frozen and useless.

Legions of geese and ducks, pulled by a cryptic call within
them, flew southward above us as we approached Lock 10. Buttons
was there, head and ears muffed against the cold, liquid steadily

streaming from his red nostrils. He was glad that the season was damn near over and he could spend his winter days on coffee and maintenance with no more towboats to lock. From his back pocket, he produced a photograph for Freddie and me.

Our grieving spirits stared out over the forming waves on Ferry Slough as I told Buttons why Smokey was not riding shotgun with me this trip.

# CHAPTER 8

# The Search for the Lock and Dam that Never Was

I suppose that Reggie McLeod chose me for this assignment because I am *Big River*'s top investigative reporter. As my son would say, "Right, Dad." I met with similar enthusiasm when I called a representative of the Corps of Engineers District Office in St. Louis. "Yeah, I kind of wondered about that once myself when I got here twenty years ago . . ." he mused, followed by a wistful silence.

Too bad. This might have been an exciting story about political corruption or an economic duel between two nearby Missouri towns, Louisiana and Clarksville (where Lock and Dam 24 does exist). Could have been about an ancient Osage curse or the poor counting skills of a corps office in the 1930s when the nine-foot-channel authorization was pushed through. Instead, it seems to just be a boring case of deciding that Lock and Dam 23 wasn't necessary.

In truth, I had once numbered this missing lock as one of my original mysteries of the river. On my first trip as a deckhand in 1973, I noticed that we locked through at 24 and, a few hours later, were at 22. However, having been previously sent back to the tow-boat for such mythical gear as ratchet stretchers and keys to locks, I was not anxious to ask any unnecessary questions. And so the opportunity evaporated as my thoughts turned to such mundane topics as "What's for supper?" and "When will my relief show up?"

Of course, the question that begs to be asked here is, "If they didn't need it, why not change 24 into 23 and so on?" Indeed, in *The Final Survey Report of the 9-Foot Channel Project*, published in January 1932 as *House Document 137*, a cost estimate of $4,842,500 was submitted for construction of Lock 23. Sometime between then and when construction began, someone decided poor old 23 just wasn't worth the money and effort. It is significant that Locks 22 and 24 are a mere twenty-eight miles apart. If 23 was there—probably around Louisiana, Missouri—it would create two very short pools. Though short pools are the norm farther north, most of the lower pools tend to be between twenty and forty miles long.

Dam 24 also has a fifteen-foot lift at pool stage, which is significantly larger than most of the dams above it. This would lead one to believe that 23 wasn't really necessary because the pool above 24 can obviously handle the extra water.

But why not just change the numbers? I suppose the answer to that would lead one to ask why there is a Lock and Dam 5 and a 5A, both built at about the same time. Or maybe they missed a lock and dam after 5, named it 5A, and took up the slack with 23, so that 24, 25, and 26 would have the right numbers.

Or perhaps the true reason will always remain one of the mysteries of the river.

This I do know: In the 1970s, when we could do such things, there were several bars close to Lock 24 for a crew that was off watch to enjoy while doing a two-hour double lockage. Had the same existed at Louisiana for the mythical 23, Pike County would have been a boomtown.

On a more majestic note, Lock 24 at Clarksville is one of the best places on the river to view eagles in the winter. Below the dam is great fishing grounds for them.

### Author's note

This story was written by me for an April Fool's edition of *Big River Magazine* in 2003 because no one seemed to know what had happened to good ol' Lock 23.

# CHAPTER 9

# Out Here, Man, a Lot of Stuff Can Go Wrong

My piloting career began in 1976 on a relic of a towboat named the *H. F. Leonard*. In fact, it was already a floating anachronism from the film-noir era when I had first served on it as a deckhand years before.

The 1970s being a halcyon decade of towboating, green pilots were minted with alarming rapidity. Companies didn't have the luxury of training a steersman for two years, as would be more appropriate for the responsibility assumed. I was "turned loose" after a mere six months, and it was not due to my brilliance. In a generous moment of candor by the boss, I was informed that, statistically speaking, I would damage a few-hundred-thousand-dollars' worth of equipment before I even began to learn what the hell I was doing. Have at it, big boy. Well, I was at zero in damages as I climbed up the pilothouse stairs for my first watch—at midnight.

I wasn't alone, though. There were two voices that accompanied me through the door that night. The first was that of my mentor, Adrian Hargrove. "Son, you'll learn more in the first ten minutes of standing your own watch than you will in any amount of time that I can train you. Why, I was never a steersman for anyone. Learned it all on my own, son." He had also told me he once was so lost on a river new to him that he had to throw a cigarette in the water to see which way the current ran. I was suitably impressed at that creative methodology.

Well, it took a little more than ten minutes, more like forty or so, but old Adrian was indeed a prescient teacher. It was a sound quite like the air being slowly deflated from a huge beach ball that broke the hazy, sticky Ohio River night. I didn't immediately recognize the sound, but I was suddenly overcome with the visceral feeling that it wasn't a good one. A rapid, excited call from the engineer on watch confirmed my fear. We had lost our air compressor. Through a series of H-valves and rubber lines, the compressor supplies vital air to the diesel engines. I soon realized that I was adrift, rudderless and powerless, in a river of confusion.

Then the other voice started talking to me. It was my mom. "Lee, nothing good ever happens after midnight," she would lecture me as a teenager when I headed out the door. I was now a living testament to Mom's prophetic nature. A moonlight mishap, floating down the Ohio at Rosiclare with my two barges in front—and my thumb up my ass!

George Bernard Shaw: "What is life but a series of inspired follies?" Lee Hendrix: "There's a guardian angel out there for all the follies of a fool." Luckily, I was soon to find mine.

A third voice, this one less ethereal than the other two, broke the silence. "The *Harlee Branch Jr.* to the *H.F. Leonard.* Are you havin' some kind of problem, Cap?" The voice squawked from my VHF radio. I never found out the name of that pilot on the *Branch*, but he had soon caught me and my tow. They held us steady until the air compressor was repaired. A very Christian-like act to save a poor pilgrim!

A year or so later, I returned the favor for someone who had lost their barges that were adrift out of Lock 50, near Cave-in-Rock,

Illinois. The current was very light, and the task was easily completed. I patted myself on the back. A good motto for river pilots from an anonymous author: "Don't pat yourself on the back too much, or you'll soon need Ben-Gay for your shoulder." Subsequently, I was waiting our turn to lock below Lock 24 on the Upper Mississippi at Clarksville, Missouri, when I noticed a blob in the radar coming toward me. Nine loaded barges had been lost from the lock and were floating down the river on their own. Being an expert "barge catcher" by then, I worked my thousand-foot-long tow out in the river to try to catch them. I was fixin' to get broke from sucking eggs the hard way.

"What is life but a series of inspired follies?" And this time there was no guardian angel. The current was several times stronger than it was at Cave-in-Rock that earlier night, and the loose barges were an unruly mass, floating and twisting willy-nilly in the flow, with a collective mind of their own. My naivete soon overwhelmed me and my fifteen coal barges. I had found myself in the middle of a true heavy-metal mosh pit, attacked from all sides and unable to exit. Final Score: barges 9, Lee 0. I did learn a valuable lesson. To wit, a man with sound judgement establishes clear limits on being a Good Samaritan. No one was injured, and not too much damage was done, but I was making a decent-sized dent in that few hundred thousand dollars.

Now, Lee, be honest with yourself and the reader and go back a bit in time. I had already made a big dent in it when I first became a steersman a few years before. We were going up the Illinois River with seven loaded cement barges. My friend Billy was tutoring me. It was Christmas Eve, an unusually warm one, though there was ice in the river as we came out of Peoria Lake into the town of Chillicothe. Billy was sitting behind me on the liar's bench reading a magazine when he looked up and said, "There's a buoy missing . . ." and he had just about got "here" out of his mouth when the lone barge, *MPC #9*, in the port string broke out of its rigging and started floating back toward us.

I called the deck crew who responded quickly. However, by the time they all got out there, the decks of the barge were awash. We had found a hard bottom, and *MPC #9* had sunk quick. Luckily, the bottom of the river was close to the top so you could still walk around on its decks, but a sunk barge, loaded with cement, can instill concrete images in your psyche. Like, it may become a permanent monument to your ineptitude. Years later, pilots will come by and say, "Yep, that's where ol' Lee sank that barge one Christmas Eve. Looks right purty stickin' up thar with all that moss on it, don't it? Wonder whatever happened to that ol' boy?" Heck of a Christmas legacy.

Well, we put every pump we had on that barge, but the water was coming in faster than we could pump it out. Then the temperature started to drop or, more appropriately, plummet. I remember the deck crew going out in shirtsleeves when it had happened that afternoon, but soon it became a sho' 'nuff Illinois River Christmas Eve as a blue norther attacked us. I think the temperature got down below zero that night, and it was a constant effort to keep the pumps running so they didn't freeze up. "The Hawk" was howling, and that barge wasn't rising off the bottom. I didn't sleep for a couple of days as we sat there trying to figure out what to do. Other boats would come by, and I could visualize them shaking their heads at us, immensely proud that they didn't have me for a steersman.

Billy saw a church in Chillicothe, down the street from a Pabst Blue Ribbon sign. He must have been sorely in need of spiritual uplift because he got someone to row him to town in the skiff. It was certainly a long service because he didn't return until morning. Ice was getting thicker in the river, and rowing him back around the floating icebergs tested my skills with the skiff. The next day, divers showed up with several huge pumps and eased some of my anguish by informing me that no water had gotten into the cargo. So, the barge could be raised and repaired! I'm sure that diver got paid quite handsomely to go into the ice-filled river to patch up that barge, and we treated it gently the rest of the way up to Lemont.

Much of that story wouldn't happen anymore, though barges and boats do still sink. Oral history (where we learned a whole lot about what not to do) these days is almost completely lacking, mostly due to smart phones. We do get paid much better, though heading into town will get you fired. We are full-on prisoners on towboats, only lacking orange jumpsuits. Probably better off, certainly safer—at least, the companies think that.

People ask me what life on towboats was like in the 1970s, and I ask them if they've seen many cowboy movies.

One year when I was a mate, we were stuck below Lock 21 in Quincy due to high water. We would be there for a few days, and we made a deal with the captain that we would get a vast amount of painting done if we could all go to town that night. The pilot was going to go with us, supposedly validating the legitimacy of our excursion.

"Town" to twenty-something deckhands and thirty-something pilots very far from their wives or girlfriends means two things, one of which you are probably just fantasizing about. The other you are quite likely to get way too much of. Trouble was that town was a few miles away and on the other side of the lock. There was no available road, and locking through was not an option due to the flood. But the gates on the dam were completely out of the water. It was a thrilling ride for several of us in that fifteen-foot-long jon boat to go up through those gates that evening before sunset. A life jacket or two may even have been stuffed into the boat's flat bottom. We made it to Quincy's boat ramp and headed for the first beer joint we could find, seeking solace through our demons.

We drank, even danced and wrestled, our way up and down a few city streets as the sky became darker and our wallets became thinner. We finally settled on a place with a jukebox that played "Satin Sheets" and "Queen of the Silver Dollar" until I almost had the lyrics memorized. There is nothing more inspiring than your crew, with thirty-five-cent draft beers held on high, joining in unified chorus with the locals, "You don't have to call me darlin', darlin' / You never even call me by my name." Returning to the boat was a distant, bad reality. The pilot met a honky-tonk angel, took off with her, and was gone until the next day.

After all the bars closed, we stumbled back down to where we had left the skiff, and it was, miraculously, still there. As I looked across the rampaging river and noticed Jim (the pilot) no longer with us, I realized I would have to take the skiff back down through the dam at two o'clock in the morning, with a Wild Turkey grin plastered all over my face. "Lee, nothing good ever happens after midnight."

Luckily, the guardian angel showed up for another inspired folly. But those gates on that dam looked like a guillotine hovering over my head as I guided the jon boat down through there. Everyone was just enough lit to consider it a jolly good time. Yee-haw! The next morning, we felt like we had each gone a round with Tiger Man McCool. The painting was a bit sloppy, except for Cloyd who had promised his wife before he left Arkansas that he would do nothing sinful while on that towboat and had not accompanied us hollow-eyed heathens.

We were sometimes a little less than behaved in those days. We didn't set out to break all the covenants of Christ, but accidentally did our best to bend a few. Some of the captains were understanding, and they were away from home, too. No GPS existed to track us down as we would head up the river.

A particularly Dionysian captain we once had would call Billy and me up to the pilothouse. "Boys," he'd drawl provocatively, "I'm beginnin' to get a mite thirsty." Ever vigilant for subtle cues concerning the captain's welfare, we soon had the yawl in the river and headed to Beardstown, Havana, Hardin, Pekin, or whichever town on the Illinois River was close.

The PBR arrived back with us, and the cards and dice came out until the cook ran us out of the galley. The engineers winked at and kicked each other under the table and took everyone's money in poker. Yeah, it was fun, and some of the fun aggravated the danger. So, in the 1980s, along came drug tests and maximum security. Now, you can't even go to town to buy a newspaper. On the brighter side, you go home with more money in your pocket, and the scenery is still inspiring.

There's been an endless succession of character-building experiences that have happened to me in five decades on the river. Exploding lock lines, flying ratchets capable of decapitation, and driving rainstorms are constant, unwelcome companions. But I've never had anything happen to me like Jake Brown did on that harvest-moon night below Grand Tower Chute.

We had been assisting another towboat pull its barges, one by one, from the sandy clutches of the river. As we were putting the last one back in tow, the wolves came snarling at us from out of the night, teeth bared. They were soon chasing our scared asses. The tow had broken loose. Sparks from broken wires and ratchets flew in every direction, barges caromed and crashed off one another. We quite suddenly felt that we were down range of heavy ordinance. Brave deckhands, not knowing which way to run, sought safety and cried for Mama. I was fortunate enough to run to a barge that was still attached to a towboat. Jake Brown was not.

Yes, Jake had an adventure that night that most could only dream about. In a three-to-four-miles-per-hour current, he floated solo on a runaway barge for two hours down the Mississippi River. Mind you, this was in the days before cell phones. He couldn't even call his girl and tell her about it, so she could realize what a gallant, worldly guy she had. No, he was just one scared, lonely man, repentant to God and all those he had harmed—alone with his thoughts and ready to run to the high side.

Thankfully, the story doesn't have a tragic ending. Someone once told me that, when barges break loose, they rarely run into bridge piers, dikes, or sandbars. Only when they have a pilot misguiding them do they hit anything. The towboat *B. John Yeager* caught Jake and his runaway barge down near Cape Girardeau. When Jake got back to us later that morning, he didn't have much to say. The size of his pupils spoke for him. The next day he packed his bags and left towboating behind. I heard he became a life-insurance salesman in Poplar Bluff, and he's got one slam-bang story to tell as a deal clincher.

These kinds of things happen much less regularly now, but still often enough that there's a fleet of pressed-shirt lawyers out there specializing in marine casualties. The unpredictable nature of the river makes accidents impossible to completely avoid. Those old-timers like Adrian Hargrove passed their wisdom down through succeeding old-timers, maybe like this humble writer. Perhaps one day, I'll turn furtively to a greenhorn and say, "Son, don't ever think you've got it all figured out. Stay ready for what's around the next bend. I bet there's something there waiting for you."

# CHAPTER 10

# Jimmy Dan

Looking out over the port head of the tow, Jeff Barlowe could not see any water at all between the barges and the bank of the river. He had held his tow in too close, and the current was setting the four barges loaded with dangerous, explosive chemicals and fuel under the point and onto the shore. The Louisiana side was the best way to run the Vicksburg Bridge northbound on a big rise, and the old *Carl Edwards* would never shove up through the channel span. Oscar had told him to hold her close to the Louisiana shore, but he couldn't have meant this close. He had to pull it out of there before the barges set into the bank or the port pier. The tow was moving so slowly . . . . He had to do something! He slung the steering rudders to starboard and pulled the head of the tow out quickly from the bank and into the current. He felt brief elation upon seeing water come down between the side of the tow and the willow trees, their trunks inundated by the flooding river.

However, this relief soon turned to another concern. Ever so slowly at first and then, with lightning rapidity, the ubiquitous flow of the Mississippi began to work on the tow. It carried the *Carl Edwards* and its deadly cargo away from the bank and out toward the bridge pier on the right. He suddenly realized what was taking place and, just as quickly, understood that he was getting into deep trouble. The entire 1,100-foot length of the boat and tow moved inexorably out toward the concrete of the pier—not only that, but being in the faster current had slowed their forward speed down to nothing, and they were only one-half to three-quarters of the way through the span. He needed to steer the head of the tow back to the shore, but if he did that now, the stern barge would hit. There was no room left to steer. Sweat began to pour from his hands, and he could feel drops already running down his side. There was nothing left to do except to try to back the tow out of the bridge. It was his only hope, albeit a slim one.

He stopped both engines and set both throttles to backing—and all was silence. He looked behind him and noticed the yellow guard lights on the side of the boat clouded with smoke. The outdated, direct-reversible engines had failed to start again. It had happened before, and the chief swore they would be fixed soon, but soon may never arrive. He knew that he was going to hit the Vicksburg Bridge and there wasn't a damn thing in the world he could do about it except watch. He had to alert the crew! He began ringing bells, blowing whistles, doing anything he could to call the crew on watch and to wake those asleep in their beds.

With a dull, sickening thud, the tow smacked sideways into the battle-scarred pier. Upon impact, the black water shot up twenty feet in the heavy night air between the barges and the pier. The force of millions of cubic feet of water coming down from the mountains of West Virginia, the alluvial floodplain of Nebraska, and the lakes of northern Minnesota was pushing the head of the tow ever faster toward the middle of the river. With the stern of the tow unable to keep up with the head, the current had maximum levering power. Soon, the barges and the *Carl Edwards* would be lying completely crossways and pinned against the piers that defined the smaller spans of the bridge—pinned there by the relentless power of the river. Slowly, at first sublimely, came the odor. The rupture in the side of the barge must have gone all the way through to the cargo hold, and its icy, deadly contents found their way into the air and back toward the boat: toxic chemicals that would either explode or overwhelm whoever was unfortunate enough to breathe them. They had to break away from the barges, but the boat was held fast to them by several strands of inch-thick steel cables.

Where was the crew? Why had the deckhands not responded to the emergency whistles? Frantically, he blew the whistles again and rang the general alarm—but still no response. By this time, the force of the river pushed the head of the tow with frightening speed on its collision course with one of the bridge piers out toward the middle. He knew the sparks created by steel and concrete coming together with such impact would likely ignite the fumes of the deadly torch this tow had become. Was everyone else on the boat already dead? As the lead barge slammed into the pier, he ducked. The force of the ensuing explosion filled the pilothouse with flying glass and debris. He pulled his body up from the floor and saw a wall of flames where the night sky had been.

The burning cold of anhydrous ammonia that had been stored at a temperature of negative two hundred degrees Fahrenheit came unabated into the pilothouse. He tried to cover his nose and mouth, first with a handkerchief and then his shirt, exposing the skin of his stomach. Too terrified to look at his flesh turning color from the searing chemicals, he continued trying to call the crew. His thoughts flashed to Debbie, Petey, and Krista—sleeping in their home in Tennessee, so far away. Would he ever see his wife and kids again? If they could get the boat away from the barges, perhaps he could still survive. Then he remembered the untimely malfunction of the engines. Yes, he was trapped, bound to an icy tomb of ammonia. Jeff Barlowe realized that he would die on this lonely, flooded river and that his wife and children may never even see his body. Only if the river decided to give him up.

"Wake up, Cap. It's 11:30. Time to go on watch. Want me ta turn your light on for ya?"

"Yeah, Yeah! . . . Hell, yes!" Jeff almost screamed, sitting up quickly.

"Gosh, Jeff, what you been doin' ta that bed o' yours?" the voice inquired, flipping on the overhead light.

Jeff rolled his eyes toward the voice and then back to his fitted sheet, which had been pulled up off the mattress of his bed on two corners, exposing the patterned print. The mattress was wet with perspiration, and both pillows were on the floor next to him. "It's time to get up?" he replied, hopefully.

"Yep, Jeff," Bobby mused. "Time to stand your first pilot's watch. Ev'rybody says they gonna go ta sleep with their life jackets on!"

"Okay, okay. I'll be up in a minute." He was still not completely certain that this scene was his reality.

Bobby left the room, bored and disappointed at not receiving more of a retort. Jeff sat on the edge of his bed, staring at the neat, tidy pilot's room around him. He rubbed his eyes, trying to straighten all the pieces out in his brain. "Wow, what a dream," he whispered to himself.

He swung out of bed and struggled to the sink to throw cold water in his face. The chill of the water brought some clarity. As he dried his face with his hand towel, he looked about the pilot's room. A feeling of pride began to swell in his chest. He had worked for ten years on the deck, as a mate and, finally, as a steersman, for the day when he would be called a pilot, sit at the head of the table in the galley, and have this private, upstairs room, not to mention a large increase in pay.

The room itself was furnished quite simply with a single bed, a small desk, and chest of drawers. All the furniture had undoubtedly been purchased from a discount outlet, but it was on the second deck, away from the noise of the engines, and he only had to share the bathroom with one other person—Oscar, the captain. "Yeah, Oscar," he sleepily said to himself. "Don't stand there daydreaming all night. Get up and relieve that man!"

While putting on his blue jeans, T-shirt, and sneakers, he thought about what his first full watch would be like. They had gone through St. Francisville the previous afternoon, where Al had gotten off, and were making about three or four miles per hour. Jeff figured that they should be somewhere below Hog Point. He then began to mentally calculate how many miles it was up to Vicksburg and if he would catch the bridge on his watch. Sensations of loneliness and uncertainty began to penetrate his pride as he left the room and headed into the dim, austere hallway. A small black and white plastic sign on the first door announced that this was the bottom of the stairs leading to the pilothouse.

He opened the door and headed up into total blackness. Thirteen steps up to the top. About halfway up, he could smell the smoke from Oscar's cigars mixed in with the ever-present aroma of freshly brewed coffee. The only sounds to be heard, other than Oscar's off-key accompaniment to George Jones, was the occasional static-ridden voice coming from Channels 13 and 16 on the marine radios. As he gained the summit, his pupils had already begun to dilate, and he could make out the forms of the radar, the steering levers, and Oscar's short, pudgy body in the pilot's chair. A sudden burst of light violated his eyes, and he squinted uncomfortably as Oscar flipped on the switch for the pilothouse lights.

"Let's have a look at this rookie pilot and see if he's ready," Oscar laughed.

Trying nobly to smile but still squinting from the intrusion of light, Jeff eased over to the coffee pot to pour his first cup. "It's all right, Cap, I don't need any light to pour my coffee," he replied.

Relief came as Oscar switched the light back off, and his eyes renewed the process of adjustment. "I had Bobby make ya some special coffee tonight, some o' that Cajun stuff with chicory," Oscar said. "Guar—ahnteed ta keep a young pilot awake between Baton Rouge and Natchez. That boy makes damn good coffee. I reckon we'll keep 'im around another trip or two unless he decides ta head back home and work in a sawmill. You know where ya's at, Cap?"

Jeff stared into the Louisiana darkness. The night was murky and humid. The water, the moss-covered cypress and willow trees on the bank, and any buoys remaining in the raising river had all been transformed into an indistinct, soupy gruel by the June haze. Jeff saw, vaguely in the distance, the ghostly beacon of two navigation lights. The closer of the two flashed twice every five seconds, and the furthest flashed once, telling him that they would have to cross the river between the two lights, if he wanted to stay in the main channel. He looked at the radar screen, mentally transposing the picture into what he knew about the shape of the river. Knowing where they had been at 6 p.m. and their approximate speed, he was ready to make an educated guess to Oscar's question. Then he noticed the outline of a bluff, the only significant height-of-land feature within twenty miles. It's cutout form reaffirmed his conjecture. "Tunica Bluff," he responded definitively.

"You got your good eye on tonight, Cap," Oscar returned. "You should have a pretty easy time of it tonight. I hadn't heard a soul talkin' on the radio. I met a coupla boats down around Morganza and don't know nuthin' else. I kinda think most of the southbounders won't be a-runnin' tonight 'cause she's startin' ta git a bit foggy out there. Now, when ya git up there ta Angola, jus' run her an easy distance off that east bank. There's good water off that point. It'll save you some time.

"An' don't pick up no hitchhikers aroun' that prison ferry."

"No sir, I won't." Feeling mentally and physically comfortable with the situation, it was time for Jeff to go to work. "I got ya, Cap," he proudly announced.

Oscar slowly rose from the pilot's chair and, with all the care and caution of a mother handing her infant to the waiting arms of the father, gave the controls of the *Carl Edwards* to Jeff. Jeff positioned his frame in the swiveled, high-backed pilot's chair and went about the business of arranging his charts, radios, and coffee cup as he wanted to have them. He knew that Oscar would stay up and talk to him for at least ten minutes before going to bed. Oscar's life was the river, and though he claimed that he could hardly wait on the next two years so he could retire, Jeff knew that retirement would not sit well with him. He speculated that Oscar would find an excuse to stay out on boats until he was dragged off feet first. Part of Oscar's conversation would be geared to helping Jeff and part of it to satisfy his own mind that Jeff was ready. Though outwardly calm, Oscar's insides would be gnashing all night, and sleep would be uneasy with the young pilot at the sticks.

"I got some river stages tonight," Oscar continued. "Forty-five feet an' rising at Vicksburg, thirty-eight an' rising at Baton Rouge, an' forty an' rising at Memphis. They must've had 'em some hellacious rains up north. Now, a man can run over most of the sandbars and some dikes ta stay in slack water. The only places we'll slow down too much is 'round the bridges. It'll be tough shovin' up through some of 'em 'cause the water'll funnel down between the piers. They really loaded our wagon with that extra barge out there. Now, when ya get up to that Old River Control Structure, you be sure ta stay on that east bank. There's gonna be a terr'ble amount o' water gushin' down through that thing. Check for traffic 'fore ya get there, an' if somebody's a-comin', you be sure an' meet 'em on one whistle. If we get this ol' boat over halfway toward that west bank, she'll suck us right down that hole, and that's one hole we don' wanna find out about." Waiting for a response from Jeff and receiving none, he asked, "Did ya hear me, Cap?"

Earth to Jeff! He finally realized that Oscar was looking for reassurance. However, Jeff was still going over the nightmare of Vicksburg in his mind. "Yeah, I heard ya, Oscar," he replied. "Sorry, I was thinkin' 'bout somethin' else. When you reckon we'll get to Vicksburg?"

"Well, we been averagin' 'tween three and four. 'Course, we'll slow down some 'bove Natchez. Vicksburg's about 437. Hey, you got a date in Vicksburg or somethin'?"

"Naw. I was just wonderin'. How do you run that Vicksburg Bridge at this stage?"

"Hadn't me an' Al tol' you 'bout that before? Only one way to run it. Gotta run that first full span out off that Louisiana shore an' hold her in close ta that point. 'Cause if'n you don't, she'll come outta there like a scalded dog, an' you'll have yourself a river full of problems, an' you'll wisht you was in Siberia or anyplace else. Say, Vicksburg's a hunnerd an' fifty mile from here. These bridges is bad enough when you catch 'em, without worryin' 'bout 'em for two days. If you catch Vicksburg on your watch, I'll get up an' help ya. Now, ya better check for traffic before ya get up ta Hog Point. An', if ya don't get this ol' boat out more in that slack water, we'll all die of old age 'fore we even make it ta Vicksburg."

Oscar's advice brought Jeff back to the stretch of river at hand. He did not want Oscar to lose his faith in him or to get up off watch to help him through a tight place. If he was going to be a pilot, he would have to make his own decisions and break the cycle of reliance he had on Oscar and Al. Of course, he would always be open to a little friendly advice. "Okay, Oscar," he smiled in the darkness. "You say run close to that point at Angola and stay out wide at the Control Structure. I don't think I'll have any problems."

"All right, Cap," Oscar yawned. "I'm goin' ta bed. Ya know where ta find me if ya need me. That's my job." He paused at the top of the stairs in reflection. "Oh yeah," he remembered. "With this big rise, there's a heap o' drift in the river, big trees an' all. You might get some hung up in them flankin' rudders, an' if that happens, this ol' tub'll commence ta shakin' an' wallerin' like a dog tryin' ta shit a peach seed. The only way ta get it out is ta back up on 'er a few seconds an' wave them rudders 'round. I hadda do it twice tonight. Chief says he'll work on that exhaust problem when we get to Lemont. Well, I'm sleepy, Cap. Drive safe."

"Goodnight, Oscar. See ya in the mornin'." Jeff heard the footsteps slowly going down the stairs, and the door at the bottom coming open. The door remained open for several seconds before finally closing.

Jeff breathed a sigh of relief. He was finally on his own in the pilothouse of the *Carl Edwards*. The boat was about thirty years old, sorely lacking in power (only about two thousand horsepower) and, at times, unreliable. However, it was a place to start, and he knew of many pilots who had started on boats like this one and eventually worked their way up to larger, newer ones—boats of 5,600, 7,200, even 10,500 horsepower, pushing tows of forty or more barges. If he didn't have accidents, paid attention to the river, and proved his independence to the company, he could do the same. An old klunker like this was all right for Al and Oscar; they had both worked on big boats earlier in their careers and were now happy with the smaller tows that the *Carl Edwards* handled. They were willing to put up with the aggravation of engines that wouldn't start due to a confounding exhaust buildup that kept the engineers scratching their heads or running out of the engine room when it filled up with smoke. There was a myriad of other problems as well, such as circuits overloading, rudder cables breaking, and a general lack of confidence in the boat.

He glanced about his antiquated surroundings and was forced to admit the boat held a certain charm that newer vessels seemed to lack. On the console in front of him was a large, round, brass cylinder. Written on the cylinder were engine commands, such as "Full Ahead," "Half Astern," and "All Stop." The cylinder had a handle on it which slid to the desired position, indicating to the engineer the needs of the pilot. This "engine-order telegraph" was no longer functional but was a classic remnant of the days when the pilot had to ring to the engineer to control engine direction and speed. At his left side was a black telephone, which was still functional and used to talk directly to the engineer. It was powered by the sound of voice, and Jeff had to constantly yell at the top of his lungs to communicate with Shorty, the engineer on his watch. Present, as well, in the pilothouse were certain tributes to modernity, like radar, a sonar depth finder, and single-sideband radio. These improvements almost seemed anomalous to the musty tone of the room.

As his six-hour watch wore on, Jeff became more and more comfortable with the handling of the boat. Though visibility remained murky, it was not getting any worse, and in the wide, deep river, running by radar was satisfactory. The depth finder never dropped below forty feet all the way up through Miles Bar, Shreve's Bar, and up toward Fort Adams. More importantly, he had removed the nightmare from his mind, and Vicksburg seemed a thousand miles away. The marine radios were quiet, just a few distant voices from way down around St. Francisville and some stray signals from the Red and Atchafalaya. "Oscar was right," he thought. "The southbound boats are stopped, and I'll have the whole river to myself."

This is one of the loneliest stretches of the Mississippi between Cairo and New Orleans. Few homes or even lights from distant shacks can be seen from the river between St. Francisville and Natchez. There were many small oil fields pumping in the area, and the odor of crude, especially on a thick, viscous night like it was, permeated the air. The only breaks in the otherwise solid gloom he would have were the navigation lights, which were dimly seen in the haze, and the lights around the Old River Control Structure.

The Old River Control Structure is designed to divert 30 percent of the river's flow away from New Orleans and Baton Rouge and into the Old River and the Atchafalaya. With this rapidly rising river stage, the gates of the structure would be allowing hundreds of thousands of cubic feet per second to pass through the rollers.

As the night wore on, Jeff and the *Carl Edwards* settled into the beautiful, rhythmic relationship that humans can feel when they become one with a machine. A handy country music station, a pleasant conversation with T. R. (the deckhand on his watch), and the majesty of the river created a feeling of peace in Jeff as the boat and he tranquilly slipped ever deeper into, and became part of, the night. Though he missed his wife and family, he knew that they understood, and he felt no remorse that, at this moment, there was no place in the world he would rather be.

As he got up to Fort Adams, he shined his pilothouse flashlight at the clock on the wall. Ten minutes to four. In another hour and a half, he would complete his first full watch in the pilothouse. Al and Oscar had frequently left him alone during his steersman's training, had even allowed him to get into a bind or two, but he had known that they were closely monitoring the situation and he would hear their footsteps on the stairs before things got out of hand. He hoped the captain was sleeping like a baby, trusting that his "partner" Jeff was capable of handling any situation that might arise. Jeff's mind wandered to Vicksburg. "Will we get there on my watch?" he thought. He began to count the miles and divide by the average speed of the boat, then concluded it to be a fruitless endeavor. "If I catch it, I just catch it, and I'll deal with it," he resolved. "And come hell or high water, I won't wake up the captain. Case closed." Ten minutes later, he was adding up the miles again.

Returning to the job at hand, he recalled Oscar's instructions to check for traffic while still below the Control Structure. "WZN 4332, the *Carl Edwards*, checking for any southbound traffic around Black Hawk Bend," he announced into the microphone of his marine radio. Waiting for several seconds and hearing no response, he gave his position on the air. Again, no one answered. "Speak now or forever hold your peace," he sighed to himself. Not surprised that there was no reply, he widened his tow out toward the east bank as he had been instructed and continued on.

Out here, the water was slack, and they were making good speed. There was sometimes a string of red buoys up through there, but at this stage of water, a pilot could cheat all the way to the willow-banked tree line. He flipped on his newly installed xenon searchlight and shined the white spear across the river. As calm as the water was on this side, so was it raging on the other.

Even through the thick mist, he could see the frothy turbulence near the spillways to the dam. Limbs and even entire trunks of trees, barrels, fences, and all other forms of debris were swept into the concrete jaws and devoured by the rampage.

A very deep respect for this carnage created a stiffening of muscles and tightening of his body. Long minutes later as the *Carl Edwards* finally crawled above the level of the dam, he breathed deeply. Glancing down at his depth finder, he thought he noticed that he was beginning to come into shallower water. Not wanting to take any chances with a chemical tow, he decided to steer back toward the main channel. That's when the boat started to shake. He remembered what Oscar had told him about drift in the flanking rudders. Concurrently, the sound-powered phone from the engine room began ringing.

The voice at the other end of the phone was barely audible and seemed as though it was coming from the end of a long, underground tunnel. "Yeah, Shorty, I know I got somethin' hung up in the rudders. You say it sounds like a buoy?" Jeff replied to the voice. "Okay, I'm gonna hit 'er a lick back." He pulled both throttles to the stop position. There were no clutches on these engines, so the entire engine had to be stopped and restarted in reverse. He waited for a few seconds, then set both controls to the astern position. Both of the direct-reversible engines sputtered momentarily before finally starting. He then waved the rudder controls back and forth twice. "Did that get it out?" he screamed into the phone after a few seconds of backing.

"No, Cap, you'll have to try it again," replied Shorty. His voice began to sound strained.

Jeff uttered a curse, knowing that he would have to repeat the entire process. That meant he would have to stop both engines, start them while set ahead, stop them, restart them in reverse, and stop them yet again before being able to continue up the river—provided, of course, that the next attempt was even successful. It was more than just a dreadful nuisance; he became aware that the boat was beginning to drift slowly down the river with the current and that the head of the tow was swinging out toward the middle. The drift had to be cleared soon. For the first time in more than an hour, Jeff got up out of his chair. "Keep me posted on what's going on down there!" he shouted. He knocked his coffee cup and charts to the floor, the cup breaking in a hundred pieces as he attempted, again, to purge the underside of the *Carl Edwards*. By the time he got both engines ready for reverse again, the tow was nearing the center of the river. Not only was the current much swifter, taking them rapidly downstream, but they were getting into heavier drift as well. He could feel his knees crashing together as he set both engines to backing.

"I think you got 'er that time, Cap!" Jeff was elated to hear Shorty's voice. The illumination from the yellow guard lights on the side of the boat revealed the flushed-out culprit—a red buoy, broken from its anchor chain and set adrift in the high water.

"Now, I've got to get her going back up the river," he whispered to himself through clenched teeth. He turned his head quickly, allowing himself a momentary glance at the dam lurking behind him. He put both controls to ahead and attempted to restart the engines. Again, the engines sputtered—and sputtered. The black phone rang quickly two times and stopped. The clouds of smoke billowing up from the engine room let him know that answering the phone was a useless enterprise. There would be no engineer at the other end, the fumes of the backfiring valves having chased Shorty out on the deck. It would be at least a couple of minutes before the exhaust would clear and the engines could be started. They were now helplessly floating in the main current toward the dam, and Jeff recognized the aftertaste in the back of his throat. "Wake me up, Bobby," he pleaded.

Even if both engines started on the next try, it was doubtful that the overloaded boat would be able to shove out of this draw. That would leave him with the frightful possibility of turning the barges loose in order to save the boat and the crew. This was a decision he could not make on his own. He prepared to ring the whistle in the galley to alert the crew and to wake up Oscar, if he was not already on his way.

"The *Jimmy Dan* to the *Carl Edwards*."

Thinking that fear had consumed him completely, Jeff initially discounted the scratchy voice coming from Channel 13 as a hallucination. However, as the hail came clearly again, he picked up the mic of his radio with his wet and shaking hand. He looked out of the windows of the pilothouse and was astounded to see a towboat without barges. He was sure there were no other boats around for many miles. "Well, if this is a dream, it's a damn site better than the last one," he thought to himself as he returned the hail. "The *Carl Edwards* back to the station calling."

"The *Jimmy Dan*, Cappy, the *Jimmy Dan*. It looks like you might be havin' a bit of trouble. Can we help you out somehow?" Though the boat was virtually right next to him, the radio transmission was extremely weak, as if the sending unit were quite old or in need of repair.

"Skipper, you sure can," Jeff returned, struggling to keep his voice from wavering. "I lost both engines trying to clear drift from underneath my boat."

"Cappy, let me get a good towing line on the side of your barges, and let's see if we can get you out of the current."

The black phone rang again. It was Shorty telling him to give the engines another try. Uttering a prayer, he flipped the throttles into the notch and pushed forward. Both engines sputtered erratically—then finally started! "Please, Shorty, give me an overload. All you've got!" he yelled urgently. Maybe, just maybe, with an overload and the help of the *Jimmy Dan*, wherever she had come from, they could get back across the river to safety. He was unfamiliar with this vessel, not certain how much power she had. It suddenly occurred to him that he had not alerted the crew, besides Shorty, to any of this. "Well, let's see what happens here before we call Oscar," he said quietly to himself.

Shorty must have known that they were in some kind of trouble because he had the engines clunking louder than Jeff had ever before heard them. Still, however, they were floating steadily backwards, soon not more than a couple hundred yards or so above the spillway to the dam. As soon as the *Jimmy Dan* got alongside, they would have one shot at getting the tow out of there. If that failed, they would be in the clutches of the river, headed toward oblivion.

The *Jimmy Dan* maneuvered smoothly alongside the barges and attached a line. The skills of their pilot and deck crew were well honed, as no energy was lost or motion wasted in the process. Silently, the pilot swung the boat into its maximum pushing position and came ahead on both engines. Almost imperceptibly at first, the sternway was halted. Jeff allowed himself to momentarily look away from the impending doom and toward the *Jimmy Dan*. She was of a style he had never before encountered, except from old photos in the *Waterways Journal*. He judged the vessel to be at least ten years older than the *Carl Edwards*; she was 185 feet long if she was an inch, and the pilothouse was set well back from the bow. She had a monstrous head deck covered by a canvas awning and wide walkways on her two bottom-most decks.

For the first time since the ordeal began, Jeff flopped into the pilot's chair as the combined strength of the two venerable vessels began to push the tow up the river. The muscles in his neck and back felt as though they had been released from the jaws of a vice that had twisted them in a knot. Taking his binoculars in hand, he took a closer look at the crew of the *Jimmy Dan*. Most were standing on the head deck smoking cigarettes. They wore dark-colored, baggy work pants and shirts with two large pockets in the front, almost like uniforms. They were apparently not encumbered by safety consciousness because no one was wearing a life jacket. However, the thing that made them distinctive among all the deck crews that Jeff had seen on the river was their close-trimmed locks and lack of facial hair. How strange it was to see a deck crew without a single member sporting a beard or moustache.

Once they had gotten the tow out of the main draw from the dam, it was a simple matter to ferry across the river to hit the slack water again. Immediately upon reaching this point, the crew of the *Jimmy Dan* turned their boat loose from the tow. Wishing to express his gratitude to his rescuers before they left, Jeff hurriedly called the *Jimmy Dan* on Channel 13. Receiving no reply, he tried on Channel 16—still, no luck. As the *Jimmy Dan* slowly crept by in the mist and headed down the river, the light in her pilothouse came on. Jeff returned the traditional towboat salute, never getting more than a brief glimpse of the pilot to whom he owed so much.

Jeff glanced at the pilothouse clock: 4:45 a.m. The entire incident had lasted about a half-hour. As he swept up the broken coffee cup off the floor, he tried to compose himself. "If I live to pilot fifty years, I'll never forget this night."

A little after five o'clock, T. R. came up to make a fresh pot of coffee for Oscar. "What the hell was all that backin' and bangin' goin' on?" he asked.

"Just clearin' some drift from underneath," Jeff innocently replied.

"'Bout skeered that poor ol' cook halfway ta death. She's a-wantin' ta know if you're comin' down fer brefust."

"Yeah. I'll be down." Jeff smiled as T. R. bid farewell and headed down the stairs. No one on the boat knew what had happened except Shorty, and he only knew part of the story. He would be able to tell Oscar the whole saga himself, doing his own editing, of course. All in all, he had performed quite admirably, and he was proud that he had not awakened Oscar and alarmed the crew. He put the broom and dustpan back behind the liar's bench and cleaned things up, knowing Oscar would be up before 5:30.

The blackness of night gradually gave way to a dripping, gray dawn as Jeff heard the pilothouse door open and the footsteps on the stairs. "Slow, but sure, the old man's comin'," he heard Oscar puff. He paused at the top of the stairs to catch his breath before heading to the coffee pot.

"Did you get your rest last night, Cap?" Jeff asked.

Oscar took a deep, reflective breath as he looked out over the river. "Man, I must have a clear conscience, 'cause I slept like a log. They had ta call me twice this mawnin' 'fore I rose up. That ain't usual fer me. How'd the first watch go las' night, Cappy?"

"Why don't you pour yourself a cup an' I'll tell ya all 'bout it." Jeff had watched master storytellers in the pilothouse weave their tales, and he wanted to give this rendering its full impact.

Oscar quietly celebrated the birth of a new river morning as he stirred the cream and sugar around in his coffee. "Black Hawk Bend," he said, correctly identifying their current position, merely saving Jeff the courtesy. "There was a boat sunk up around here—leastways, they think it got sunk, nobody's quite fer sure." Jeff obviously would have to put up with one of Oscar's old stories before he could tell his true-to-life one. It sat well, however. He had the time.

"They're not sure?" Jeff asked, trying to sound more interested than he was.

"Yeah. Boat left outta Baton Rouge, the *Ginnie D.*, I think it was. Well, anyway, this was quite a while back—I was even young. Like I say, they was headed for Vicksburg ta pick up a tow, an' a few days passed, an' they never showed up there. Well, radios in them days wasn't no 'count, an' the las' anybody heard from 'em was down 'round St. Francisville. They sent boats out ta search fer 'em, even the Corps of Engineers got into it; most of 'em think they sunk into one o' these here deep spots and got theirselves washed deeper an' deeper into the bottom. There wasn't no divers back in them days that'd come up here an' look fer 'em. This water's over a hunnerd feet deep in places, ya know." Oscar seemed immersed in his thoughts. "*Ginnie D.* wasn't the name o' that damn boat. It's hell ta be a-gittin' old an' your memory don't serve ya no more. Well, what happened with you las' night? You actin' like ya got a story ta tell."

Leaning back in the chair, Jeff began his account. "Well, she was pretty dark last night, like the inside of a cow. And I was comin' up past that Old River Control Structure, out wide, jus' like you tol' me, and then—"

Oscar slapped the coffee table emphatically. "*Jimmy Dan*, by God! . . . He looked sheepishly at Jeff. "Sorry, Cap, I didn't mean ta interrupt you. Go on with your story."

"I was runnin' in forty feet of water . . . What did you say?" Jeff stopped suddenly.

"The *Jimmy Dan*. That was the name o' that ol' ghost boat."

"Ghost boat?"

"Yeah, that one I was a-tellin' you about. Some of 'em says there was somethin' strange 'bout the crew on her. Said there was some murderers, thieves, and the like. They didn't check no references on crews much back in them days. They mos'ly took whoever they could get. Well, some of 'em says a murderer got on, or maybe two, an' kilt most of the crew, an' the boat sunk on a snag or an ol' wreck or somethin'. I got ya, Cap," he announced, moving in to take over navigation for Jeff. "Well, there's some pilots even says they done seen that damn ol' boat a-floatin' 'round out here at night. Mos'ly pilots got too many days in, if you was ta ask me. Ain't that silly? You gonna let me have it, or you wanna work all day, too? If'n ya don't git up outta that chair, I'll jist go back ta sleep."

Jeff moved silently to the side, relinquishing the controls. "Well," Oscar sighed, "let's hear about you las' night. I'm tired o' my ol' stories."

"Oh, just got some drift hung up in the rudders, that's all," he said feebly.

"And . . ."

"That's all," Jeff peeped.

"That's all?" Oscar asked incredulously. "You built me up for that." He shook his head. "If that don't beat all! We kin teach young cubs the river, but we cain't teach 'em to tell a good lie or two. So, did you git the drift out?"

"Yeah. Backed on 'er a coupla times," he replied, staring out the window.

"Well, hallelujah! If that's all I kin git outta ya, you best go down an' git in that bed. They'll be a-callin' your ass about 11:15 or so. Any other boats 'round this neighborhood?"

"Haven't heard a thing all night, Oscar," he said, finding the stairs.

"Say, you ain't still frettin' 'bout Vicksburg, are ya?"

"Vicksburg? Where the hell is that?"

### Author's note

When I started piloting, I was often awakened, drenched in sweat, by dreams about what the river could do to me and my tow. These nocturnal demons, born from the power of the Lower Mississippi, in particular, could cruelly keep me from getting the sleep that a pilot desperately needs. In conversations with some other young pilots, I also learned about the fears they lived with. I was inspired to write this story set in the 1970s because of those discussions and an old legend I read in B. A. Botkin's "A Treasury of Mississippi River Folklore," entitled "The Phantom Steamer of Raccourci Bend." The erstwhile Raccourci Bend is but a few miles from the present Old River Control Structure.

# CHAPTER 11

# Sounding Machine

Ain't you got no family, no place to be
Out on the river on Christmas Eve
—John Hartford, "On Christmas Eve"

My deck crew on the *M/V Mississippi* surprised me one day, in 2010, by producing a lead line out of the deck locker. The Corps of Engineers, as it should, has everything. A lead line is an arcane instrument left over from steamboat days for sounding water depth. A heavy piece of metal (hence, "lead") was attached to a line and allowed to sink to the bottom of the river. Along the line, different materials of assorted colors were attached for the "leadsman" to see and feel, then relay depth of water up to the pilot.

Almost anyone who has read anything about river history knows that Samuel Clemens took the nom de plume "Mark Twain" from a lead-line reading, indicating two fathoms or twelve feet of river depth. Today's modern towboats, of course, have sophisticated, but much less romantic, gadgets that electronically supply that information. Some of these are known as "transducers." They give a pilot the instant gratification of knowing the depth of the water immediately (in tenths of a foot if you want). To receive this information, the navigator does not need to open his pilothouse window on a cold night and listen for the sepulchral song of the leadsman coming from the forecastle. No, it is all in a neat and tidy black box: no muss, no fuss, and, the bottom line, less people to employ—less payroll. Hooray for technology!

Most of my first pilot gigs in the 1970s were on boats on the Illinois River. We didn't have any of these transducers, and I didn't really want one because it would have scared me half to death. If you have more than twelve or fifteen feet of water in most places on the Illinois, you might as well be in the ocean because those places are rare indeed. The bottom of the river is very close to the top from Grafton plumb up to Joliet and, after that, who cares anyway? You know it's too shallow when you see dead carp and thick gumbo mud spewing up around the side of the boat.

I was pretty content over on the Illinois and could have spent the rest of my career there. But these were the 1970s, a decade of rapid industry expansion, and pilots needed to know how to run everywhere. My company insisted I train to run the Lower Mississippi River. They expected me to become a "heavy tow man" on a Lower River boat.

This 5,600-horsepower boat they put me on naturally had one of those transducers staring straight in my eyeballs where I couldn't avoid it when I was piloting. I was informed by everyone who mattered that this little piece of equipment was something you never wanted to ignore. Lower River pilots were always switching over to other channels on the radio and saying things to each other like, "I come right up that Arkansas shore, and the least water I got was twenty foot" or "You kin steer that place; there's seventeen foot on that red buoy line."

The "ducer" we had on this boat was an interesting setup: a video screen with a vertical scale from one to thirty. A horizontal squiggle would intersect this scale indicating how much water was underneath your tow. A rather heavy metal pole with a "foot" on the business end was dropped off the head of the tow, a thousand feet from the boat. It stuck in the water about a foot or so, and electrical cord was run back to the boat, the signal thus transmitted up to the pilothouse.

Very often, there is much more than thirty feet of water depth on the Lower—sometimes it is a hundred feet deep or more. If you were in deep water, you could flip a switch, and the scale would appear in its "times-ten mode," which simply meant that, if the squiggle appeared at the five-foot mark, you actually had fifty feet of water.

Our company was well aware that sending a greenhorn like me down the river to New Orleans with twenty-five loaded barges was risky business, so on southbound trips, they would provide an assistant port captain (APC) to ride with me. His job was to "keep that boy outta trouble." It may have been a logical theory except for the fact that the APC they usually sent didn't know much more about the Lower Mississippi River than I did. He was a good pilot and a fine fella, but he was as much out of his league as I was out of mine. I guess they just figured two incompetents had to somehow be better than one.

His name was Murph Parker, and even though he was pretty much just a traveler and I received less pay when he was riding down with me, I enjoyed his companionship. He had spent nearly his entire life on the river, knew everybody worth knowing, and had an inexhaustible repertoire of stories to tell. He was friendly, didn't smoke, and made good coffee. Murph did have one habit, though, that was somewhat disconcerting. He always carried a large key ring along with a pocketful of change. Anytime he was nervous about someplace on the river, he would stick his hand in his pocket and begin to rattle things around a bit. The more nervous he was, the more that change and keys would clatter.

Murph and I came on watch below Greenville, Mississippi, on the night before Christmas. We were getting well down into

the "land of the Delta Blues," and if you don't have the blues on
Christmas Eve on a towboat, you may not even be capable of feeling
loneliness. Towns down there are fifty miles or more apart, and
you can't even see them because they're behind levees that keep
the river from its natural bed that folks now call Main Street. It
was not at all like good ol' Joliet, where you could always look in
somebody's living room at 3 a.m. if you were lonely. The river is
so wide that, on a dark night like this one was, you often felt you
were free-falling into a bottomless black well. In certain places,
however, the river pinches down into a tight bend where the water
runs like a mill stream.

We had one of these "bad places" on the agenda this moonless
Christmas Eve. The river makes a hard left turn near Lake Providence, Louisiana. It is such a tight, sharp turn that, on previous
trips down, we had been "flanking" it. It always caused me grave
consternation when I had to do that.

On the Lower Mississippi, the current dashes down into bends
whose banks are normally lined with concrete mats put there by
the Corps of Engineers. Pilots on big tows can sometimes run out
of "steering room" if they try to steer their massive weight around
a tight bend. Running aground on the sandbar on the inside of the
bend is bad enough, but piling a tow up on concrete revetment on
the outside is an event one could spend a career trying to forget.
Flanking tends to be more controlled and safer—if you know how
to do it. Simply put, when flanking, a pilot doesn't even offer to try
to steer around the turn but rather points at what he most fears,
the bank revetment, and reverses engines to bring the tow down
to the current's speed. The current swirls down between the bank
and the tow, washing the tow around the bend. There are other
factors at work complicating it further, but that is the principle. I
wasn't very good at it.

Dishes would rattle and jump in the galley and searchlights
would bounce around on top of the wheelhouse as I "hossed her
back." Water being forced under the hull of a boat will have that

effect. However, there was some hope that maybe Santa Claus was gonna pay me an early visit because the word was out that a rising river had opened up a "pointway." This meant we might be able to do a straight shot and not have to flank way down in the bend.

There was a padlock on the door to this house of dreams, however, which was the fact that neither Murph nor I knew exactly where this pointway was. We had gotten some "posting instructions" from other pilots over the radio, and we could probably stumble around and find it, but you can't stumble too much with 40,000 tons of beans and corn in front of you because twenty-five barges won't just steer around any old stump. We spent a good part of our watch mentally preparing for, though not always talking about, the big event, which would occur about 3:30 a.m.

Well, we got down to about two miles above where we figured the channels split, and I asked Murph what he thought we should do: flank the bend (the known evil) or try the pointway (the great mystery). And he said, "You're the captain."

Processing that response, I figured he was either an expert in empowerment psychology or was just as ignorant and scared as I was. I hoped for the former but kind of suspected the latter. I was about twenty-seven years old and, if it was up to me, I'd vote for the great mystery any time. I could see some channel-marking buoys in the radar, but they didn't seem to make much sense the way they were lying, so I switched on a searchlight. What I saw made me wish I was just about anyplace else in the world except where I was. It looked as though buoys had been dropped out of the air from 35,000 feet—they were everywhere and anywhere, forming no discernible pattern. Waylon Jennings's voice began singing to me out of the ethereal darkness of the pilothouse walls, "Are you sure Hank done it this way?"

It was too late to turn back, so for better or for worse, here we come! There were some black and red buoys that started to take shape, but they were so far apart that you just knew the channel couldn't possibly be that fat. Out in the middle of this wide expanse of water on a dark night, without a bank to look at, it was impossible to judge rate of movement. We had not hit anything yet, so I had what reliable sources call "guarded optimism" when Murph said, "What kind of water are ya gettin'?"

Oh, the transducer! We had been running in such easy, deep water all night that I had turned the silly thing off—with it on its times-ten mode—and forgot about it. So, when I reached up and flipped it on, it read nine feet. I have never been a math genius, but it don't take a degree from MIT to figure out that a pilot with barges drawing between eight and nine feet is on the slippery slope in nine feet of water.

And that's when I started to hear it, real slow at first, like a tinkling in a Salvation Army cup. I saw Murph was not sitting down, and his hand was in his pocket, and he was doing his act.

"Shallow in here," his voice dripped, none too strong, behind me.

And a metallic taste started to come up from my throat into the back of my mouth. Beads of sweat, like drops from that leaky faucet you never fixed but meant to, suddenly welled up on my armpits and palms.

And then the ducer got down to eight feet, and I figured we were less than ten seconds from coming to a screeching halt, and there was not much I could do about it but watch it happen. I guess Murph was figuring out how he would explain this one to the office, and I was thinking about what it was gonna be like to spend Christmas pulling twenty-five loads off the ground one by one 'cause we were both sure that the sparks were gonna start flying any minute. But we were way past the point of no return, and I was sure Hank never done it this way.

And then it dropped down to six feet, and those barges either had wheels on the bottom or the sand was awful soft. And the rattling in my ears was getting stronger and stronger and, brother, when that ducer hit five feet, it sounded like just about every slot machine in Las Vegas had paid off at the same damn time.

"Are we still movin'?" he said with his voice sounding like it was coming from down in the galley someplace. He was looking out the window, and sure enough, the buoys were still going by us.

"God must be with us tonight!" he said, and it's funny, but here was a guy who had never ever showed me a religious side before but was now trying every angle.

And when it got down to three feet, I was a believer in this new order, too—until I looked down at the switches on that confounded black box and remembered what I had done—or forgot to do.

I jumped straight up in the air and shouted so you probably could have heard it in Vicksburg, "Damn, Murph, would you look at that!" as I flipped the switch over, and we were quite suddenly in over twenty feet of water. Joy to the World!

And just like that the change stopped rattling, and his face went blank, and his legs stiffened up under him, and he said, "I got to go downstairs." And then all I could hear was his footsteps on the stairs and the door closing as he left. I didn't see him again for the rest of the watch. And I shut the black box down and wished for the life of me that I had a good leadsman with a loud baritone voice instead of a sounding machine.

It was the very last time Murph ever rode down the river with me.

# CHAPTER 12

# Hubie and the *Hawk*

At the beginning of a towing season on the Upper Mississippi many years ago, we had orders to bring the tug *Black Hawk* with us up to Genoa, Wisconsin. The old *Hawk* was a comical appendage next to us, and our fifteen-barge loads of coal, appearing much like a baby piglet nuzzling up to a mother sow.

The only crew member on the *Black Hawk* was a swarthy-faced Badger named Hubie. He did not say too much and kept to himself, except to come over on the *Ruth D. Jones* with us for his meals. On these occasions, he would dive into his food with the gusto of a man who had given a fair day's work. By the time we reached Cassville, Hubie had the *Hawk* looking better than she probably had since they first dropped her in the water at Greenville years before. The only times that he would noticeably cease his labors of cleaning, "souging," scraping, and painting were when we would pass slowly by a Pabst Blue Ribbon sign on the river's bank. Then his hat would come off, revealing a reddish, balding head. He would mop his brow and longingly eye the tavern as if it were a temple.

As I came on watch one afternoon, spring was beckoning to us through the budding of maple and oak left dormant so long by the

Wisconsin winter. It was a warm day with puffy clouds floating like cotton candy balls on the pale blue face of the sky. I grabbed my first cup of coffee as we chugged past the lower power plant at Cassville. I was confident that Hubie was mesmerized by the friendly red, white, and blue sign from the Old Deniston House Hotel in town, and that he was counting the hours until we would safely deliver him and the *Black Hawk* to their home in Genoa.

The stretch from Cassville to Lock 10 at Guttenberg, Iowa, is narrow and winding, affording scant places to meet another boat. Therefore, I checked for traffic on the marine radio and, not hearing any response, figured they were either not there or not talking, so I kicked my feet up to enjoy the day. Barring any difficulty, I could probably make it at least close to Prairie du Chien before I went off watch at 6 p.m. I was about twenty-seven, taking a big boat and tow up the river, making good money for the times, and feeling quite cocky about myself. You couldn't ask for a better intro into a scene of humiliation. Pride cometh before the fall.

Soon, I was aware of another presence with me in the pilothouse. It was not unusual to see Harry's face looking at me from the top of the pilothouse stairs. Harry made beautiful models of towboats from scratch and would often stay up two or three days finishing his latest masterpiece. From the size of the aperture in his eyeballs and the vigor with which he was gnawing on his pipestem, I could see that he was ready to engage me in some abstract conversation. He soon had me babbling with him about British strategy in the Falkland Islands War, contemporaneously being waged in the furthest reaches of South America.

Somehow during my reckless prattle with Harry, attention to duty was compromised. As Harry rambled about the efficiency of British naval equipment in cold weather, my steering rhythm was broken at Island 189, and I realized that I was going much too fast and had waited way too long to steer my thousand-foot-long tow around the turn.

As I witnessed the developing situation, Harry's dialogue with me became a vacuous and rapidly diminishing monologue, a solo chord played redundantly. Somewhere in his soliloquy, he looked up to see the current take our lead barges and sweep them over the top of a submerged rock dike, wedging them well out of the channel behind Island 189. In the middle of a sentence that started off about tactical air diversions, he changed to, "Oh, shit! What did you do? How ya gonna get out of this?"

Beads of sweat that had formed like RAF planes, queueing up to take off down my arm like good English aircraft would, began to soak my shirt as I formulated a strategy for trying to pull my barges back into the channel without breaking them all to hell. Somewhere in the back of my mind, I was also formulating a plan for telling the captain, presently sleeping soundly in his bunk, how I ever got into this mess to begin with.

"You want me to go down to my room and get my camera?" Harry asked, his voice still out of sync with what was transpiring.

Unable to imagine for even a moment why I would want a souvenir photo of this scene, I flatly, quietly said, "No."

"You might want it for insurance purposes, for the salvage team when they come. This is gonna be a mess. Oh, shit, what are you gonna do?" he went on.

I was fortunate that the *Ruth D. Jones* was such a good boat because I was able to pull the barges slowly off the ground. The tricky part would come when I would have to straighten them out. Picking up some sternway, I tried to lever the tow back into the channel, but in doing so, I was forced to back my stern toward the island. The water was extremely high, and I knew that my stern would probably intrude right into the trees. I had to take that chance, however, because if I missed this time there was no telling if I could get out again. I heard the crack of timber around me as the head of the tow swung back out into the channel, and we were freed!

I told Harry to go out to check all the barges for potential damage.

"Want me to take pictures?" he asked.

Clearing my throat, which felt as if my plumbing had backed my bowels up against gravity, I merely hissed a mild negative response as he ran out of the pilothouse.

It was only then that I looked down at the poor old *Black Hawk*, still hanging along our starboard side. Hubie was standing in the middle of a scene that looked for all the world as if we had driven through a storm of branches and limbs. It seemed as if half of the timber in Wisconsin had taken residence on what had minutes before been his brightly cleaned vessel. I could imagine the scrapes, tears, bent flag poles and handrails, and broken windows. As happy as I was that real catastrophe had been averted, it was a sobering afternoon for me as I attempted to reconcile how I had ruined all this man's efforts through my negligence.

When I went down to the galley for supper, our mini adventure was the hot topic of discussion. Hubie sat quietly at the end of the table as I fielded questions between bites of corn bread and peas.

Finally, someone got right to the crux of the matter and asked me how I had managed to get myself into that situation in the first place. I saw Hubie raise his eyes in my direction. I thought of all the possible excuses—rudders hung up, tricky crosscurrents, meeting another boat, high wind—but decided that the truth was, "I just screwed up, that's all." Hubie silently rose with plate in hand and left the galley.

The next morning, we arrived in Genoa and dropped the *Black Hawk* at its home base and Hubie with it. He did not look me up to wish me a bon voyage. The whole episode passed from my life, I thought for good.

Two years later, I lived in Alma, Wisconsin, for a summer. Toward the end of August, some friends invited me to a festival down in Stoddard, about ten miles north of Genoa. It was one of those affairs that has beer tents, a women's softball tournament, four or five country bands under a circus tent that all played "Rocky Top, Tennessee," and people from around Vernon and Crawford Counties who look at La Crosse as the big city.

As I sat talking to Yank from Stoddard and tried to work up the courage to ask the shortstop from Viroqua to dance with me, I happened to look over to my right—and there was Hubie.

Well, there was no use ducking it. I had to go over and take my medicine. "Do you remember me?" I meekly asked.

Hubie, who had imbibed one or two Leinenkugels, wrinkled his brow, looked me up and down for a few seconds, and set his jaw square at me. "You're goddamn right, I remember you! You're the son of a bitch who put me up in them trees!"

As he scrutinized my face, I hoped he hadn't drunk so much beer that he wanted to take all the work it had cost him out of my hide right on the spot.

"And you know what?" he continued. "You was man enough to say that ya just screwed up, and I liked that." He wrapped his husky arm around my shoulder and led me toward the bar. "Let me buy ya a goddamn beer."

At least up in Vernon County, Wisconsin, it's a damn good idea to just admit it when you screw up.

## CHAPTER 13

# Ain't No Such Thing as a Purty Bridge

A few months back, I was honored to tell some yarns at the Lake City Arts Council in Minnesota. Afterwards, during Q and A, a lady asked me what bridge I thought was the most attractive on the river. I was left momentarily speechless due to the fact that "attractive" is not an adjective I associate with bridges, normally thinking more in terms of scary, heinous, challenging, or, better yet, ones I'd like to see blown out of the water. Historically speaking, rivermen are not usually fond of bridges since they can all hinder safe navigation. I finally offered that the new bridge in Red Wing is not too unpleasing to the eye, if you aren't coming down on it with a tow. Like all other bridges, it is much more picturesque when viewed astern.

The question did pique my mind, however. Which bridges can produce the coldest of feelings at the least welcome of times? Almost all pilots I know have a bridge that they look forward to less than others. Of course, there are many factors that go into this analysis, such as tow size, river stage, visibility, and wind speed and direction. To be sure, if you have ever hit one, or even come close, that memory will always produce a pucker factor.

Certainly, the I-20 bridge at Vicksburg, Mississippi, is high on anyone's list who has ever had to take a tow through it. There has probably never been a bridge (actually, two bridges since the old railroad bridge is immediately above it) erected in a worse place. It doesn't sit at the upper or lower end of a bend but rather lurks in Gothic wickedness square in the bight of it. If all one ever knew about the bridge was the dimensions, it would appear fairly innocuous. The main span is over eight hundred feet wide, and even heavy tows of up to forty-six barges southbound are only about one-third of that width. Plenty of room, right? Except for the fact that a large tow will come around that bend sliding down away from a significant sandbar at the mouth of the Yazoo River in low water or into an apocalyptic crosscurrent in high water. Often, above that bridge, you will be going almost as fast sideways as you are going forward. The hideous battle scars on its piers are straight out of a Tarantino movie, and not all are left there by rookies.

There exists an old story about a pilot who hit the Vicksburg Bridge with a tow and was fired by his company. He came back home, and his wife, who was counting on a big towboat paycheck, asked with alarm why he was back so soon. They lived close to Vicksburg, and he merely said, "Let's take a ride, and I'll show you." They got on the bluff looking out over the wide expanse of the river and the bridge as he solemnly recounted, "They fired me because I hit that damn bridge!"

His wife, not burdened by an overabundance of empathy, after gazing at the eight-hundred-foot-wide span, replied, "Well, I don't blame them. Anybody who could miss a hole that big oughtta get fired!" Lucky for her, she never had to make it. The current in there during high water can be seven or eight miles per hour. Countless barges and, unfortunately, some boats have been sunk on its concrete, in over a hundred feet of water, because the current sets so hard down into the bend. Sticking a tow through the Vicksburg Bridge is not for the weak of heart.

The largest tow I have personally taken down through Vicksburg is twenty-five barges. But I have always thought I might want to take a stab with a larger tow. The time for that, for me, has probably come and gone, but I did try making it on a simulator with thirty-five loads on a flood-stage river not long ago. As I descended the virtual river, the words of Captain Wamp Poe came to me: "You hold on the right pier until you get a chill—then wait another minute, until you get a fever—then wait another minute. When the mate runs out of the pilot house screaming that you are a lunatic—you wait another thirty seconds. Then you can consider pulling off the right pier."

When I started around the bend, I thought it looked pretty good, but things went downhill quickly after that. I crashed and burned. However, I am a persistent fellow when there are only animated consequences to deal with. On the seventh try, I finally made it. The man who worked in the simulator room asked me, "Well, did you finally make it?" My reply was, "No, on the sixth try, I knocked the damn thing in the river, so nobody will ever have to worry about it again." I would probably receive a slew of Christmas cards and best wishes were that only true.

Bridges on the Mississippi River are a relatively recent danger. Indigenous folks in birch-bark canoes didn't have any of them to worry about. According to MinnPost, a news organization out of Minnesota, the first bridge to cross the river was the Father Louis Hennepin Bridge above St. Anthony Falls built in 1855. In April 1856, the Chicago and Rock Island Railroad built the first rail bridge over the Mississippi from Rock Island, Illinois, to Davenport, Iowa. It was destined to become famous about a month later when a steamboat named the *Effie Afton* had an allision with it.

The owners of the *Effie Afton* sued the railroad for constructing a menace to navigation. A trial ensued in Chicago in 1857, with an Illinois lawyer named Abraham Lincoln representing the railroad.

In summation, Lincoln argued that a man has just as much right
to cross the river as another man has to go up and down it. After
several legal battles, the bridge was allowed to stand. Railroad
bridges remain as a bane to Upper Mississippi pilots, symbolic
of the bad blood between river and rail.

Some of the most hideous swing railroad bridges on the Upper
have been replaced by double-wide lift bridges, such as those at Burlington, Iowa, Hannibal, Missouri, and Hastings, Minnesota. However, those in Louisiana, Missouri, Fort Madison, Iowa, Crescent,
near Rock Island, Illinois, Sabula, Iowa, Dubuque, Iowa, LaCrosse,
Wisconsin, Pig's Eye, Minnesota, and the infamous Clinton, Iowa,
still remain. I landed so hard in the Clinton Bridge one day with
fifteen loads that the water shot into the air like Old Faithful. I only
broke twelve sets of rigging and considered myself to be lucky at
that. The old Pearl Bridge at mile forty-three on the Illinois got hit
so often one year that the railroad simply quit repairing it until a
new lift span was put in. The Illinois River has done an admirable
job of replacing notorious old bridges with new ones.

One of the more infamous and despicable bridges on the Upper
Mississippi is the Eads Bridge in St. Louis. It was designed and
built between 1867 and 1874 by the renowned riverman James
Eads. He didn't do his waterborne brethren any favors with its
placement, however. Its steel arches can kill a river pilot in high
water because you can get your tow between the piers while the
arches come right into your pilothouse with you, if you are not
close to dead center of the 520-foot span. Obviously, the higher
the water, the less clearance a lonely pilot has on those grotesque
arches. It is a historical landmark in St. Louis, but a classic example of how there are no pretty bridges as far as most river pilots
are concerned. Sitting about midway of a long, sweeping bend
through the dingy, rusty fleets and docks of St. Louis Harbor, the
bridge has a strong left-hand set. Even as historically significant
as it is, most river pilots would pay a fair sum to purchase a raffle
ticket for the opportunity to detonate it into the land of oblivion.

In fairness to the lady's question, on reflection, there are a couple of cities that have lit up their bridges in a neo-psychedelic manner. Of course, since the advent of five-panel drug tests, the complete effect of the swirling, multicolored lights cannot be fully appreciated by the approaching navigator. Memphis on the Lower Mississippi and Little Rock on the Arkansas River have invested funds in this type of eye candy.

I think that the one bridge that gets my attention most out of all the rivers we run, however, is the Simmesport Bridge on the Atchafalaya River. I had a rough start with it. We were delivering a passenger boat down to Jennings, Louisiana, for a retrofit to a casino. We didn't even have to go down the Atchafalaya, but someone decided that it was the quickest path to get there. It may have been, but we came close to sinking the boat. When the water is low, there is a fixed bridge span that boats can fit under. When the water is high, however, the bridge tender must be called out to crank the ancient swing span open. It only leaves you a little over 120 feet of horizontal clearance with a strong set to starboard.

Since all our living quarters were impromptu, I was berthed in a bathroom under the top deck. I had just gotten to sleep when I heard several loud thumps on the steel plating above me, then the sound of air escaping. This monstrosity of a bridge has a swing span without any protective cells or a sheer fence. In other words, the pilot on watch had driven the boat right into the naked steel of the trestle. Penetrating our armor like a lance would, it wiped everything out on the upper deck of the boat. Had the boat become wedged under that steel, it likely would have flipped over. At least, we were already headed for the shipyard.

To add to the confusion, not all the dangers of transiting bridges are associated with fitting between piers. On many large passenger boats, and sometimes the *M/V Mississippi*, stacks, radar scanners, and mast poles need to be lowered. If you forget this nicety, a quick metallic haircut accompanied by a loud racket can be apoplectically achieved. The *American Queen*'s stacks are so high that you have to lower them at every bridge above the I-10 in Baton Rouge. Even the pilothouse of that boat has to be retracted for most bridges above St. Louis. I used to experience screaming nightmares about coming down on bridges and forgetting to lower something when working on that boat. That's a true bummer when you only have about four hours to sleep.

The weirdest experience I ever had with a bridge, however, came at the Sabula, Iowa, Railroad Bridge one night at about 1 a.m. I was northbound with fifteen loads of coal and called the bridge tender for an opening. She told me that a train was approaching, and I would have to wait for a few minutes. Being in my twenties, I wasn't always given to wise decisions. I asked if I could stick the tow up in the 160-foot-wide span and sit there until she could open it up. With stacked covers sticking high over the cargo box of the coal loads, that would work as long as she swung the trestle upstream, which I asked her to do. I should never have stuck the tow in the bridge in the first place, then shouldn't have said anything to confuse her about which way to swing the bridge. Double fault!!

*Screech*!! A swinging steel bridge wins against fiberglass barge covers every time. At least she told me she was sorry. Of course, I still had to fill out the accident report. It was in the "cowboy" days before you had to pee in a cup after a boo-boo, so we were both spared that inconvenience.

Most pilots will have their own rogues' gallery of favorite bridges. These have been just a very few of mine.

### Author's note

Some of this essay is written with tongue in cheek. Bridges, of course, are a necessity to modern life. When I write about destroying one, I am squarely in the realm of wistful hyperbole. Some bridges have collapsed after an allision with a river craft, and that is a devastating event that no one wants. Many of the heinous bridges on the river were constructed so long ago that the builders did not envision the size of the tows that we push today.

And, of course, the Eads Bridge is a historic landmark in my birthplace of St. Louis. I understand it to be the first steel bridge constructed in the world. But it still gives me nightmares.

**CHAPTER 14**

# Connecting People with the River

In 1972, when I started on the river, many people of my age who had money were purchasing Eurail passes and going on trips to Europe. I didn't have any money. However, I met a guy who had worked on the river for a month, saved his pay, and then went to Europe. That had sounded like a potential template for me. I started by getting the job on the river, then that part of the plan stalled out. By 1983, I had eleven years in, had become a pilot, had lost a few girlfriends along the way, and still hadn't made the mythical intercontinental trek. I had managed to finally save some loot, ergo it was time for my sabbatical to begin.

I was single, with working-class tastes, and a few bucks could last me quite a while. The Reagan dollar was strong, so Europe was cheap. I didn't kiss the river goodbye but just said, "See ya, I'll be back." Then I got to Europe and found I liked it, though I had to make my money stretch further by instituting personal austerity and finding employment. One of my jobs was with an outdoor-adventure center in Germany. I liked that also.

I began formulating plans. Adventure education on the Mississippi River seemed like a good goal. When I returned to St. Louis, I was fortunate enough to meet Hank Schafermeyer. He was helping to facilitate the national Mississippi River Youth Expedition, and he felt I could be an asset. I must have done alright because he then offered me a paying job. He and a group of dedicated staff facilitated outdoor, adventure education through a public-school desegregation program known as SLEAP (because that's what you never did). It seemed as though my dreams were beginning to take some shape.

To make a long story short, my vision never materialized as I dreamt it would. If dreams ever do, that's when you have to be wary. However, I had a great deal of fun doing other satisfying stuff and meeting people who shared some of my love for the river. Coworker Tom Ball and I became river paddling partners.

I found he loved getting out on big water in small boats as much as I did. We had plenty of equipment available to us, and we used it to connect people (primarily teachers) to the river. One of the programs we facilitated was a river-cleanup organized by the Corps of Engineers (with whom I would later work for nine years) called Adopt-A-Shoreline. We picked a stretch over near Alton, Illinois, and had at it. Along with some other eager folks, we patrolled the bank and even submersed ourselves in the river digging out Styrofoam, forsaken appliances, tires, furniture, thousands of plastic bottles, and any other trash people had discarded over the years. I have never met anyone so eager to dive in the Mississippi for a good purpose as Tom Ball was.

Tom and I later organized many trips on the river near St. Louis, then decided to venture out. I had shared with everyone how different the river was much further up. We organized two trips that were each about a week long. On one of the trips, upon conclusion, I asked everyone to share with me what the week on the river signified to them. I am honored to include their writings in this book.

This river has become the blood in my veins. I feel as though I am the river. Cat's paws on its surface are the breeze on my face.

Looking out at the river sometimes get a flash of awareness of the absolutely massive scale of it.

Matter in motion across enormous distances.

And I am privy to see this one small instant of the water,

The mud,

The rock,

The wind,

As they move their way across the Earth.

The point of the present echoes back to the beginning of time.

The power of a river trip on the Mississippi, an honor, a modern-day semirepetition of an ancient way of life that seems like a vague and far away but still palatable, at times becoming of a cellular, memory. A slow, burning sense of ancestral life ways. An awakening.

The majesty of the river and the life she supports, both ecosystems and human livelihoods, the ongoing issues of controlling the raw power, commercial use versus natural flow and changes, the issues of exotic plant and animal introduction and pollution.

Often overwhelming topics don't seem to speak to me as strongly as does the history—the people and strength of their connection to and respect for the river. Ownership. Pride. Reverence. Almost unspeakable lessons . . . the sense of paddling through quiet sloughs, sleeping on sandbars, walking past ancient mounds—burial sites of early Native Americans, mud and sand squeezing between each toe.

The sheer beauty of the river.

Is there any way to push back time to help the inspirational animals and birds. The river is a complex thing.

Out of sight, out of mind.

People are always ready to be of assistance. You never know what is around the bend. Water is the source of everything. It inspires me to see that there is still life left—all the lilies and the birds.

One vast connecting network. The old and the new.

Taking things for granted. It's important to see the basis behind everything.

Continuity.

The picture of a blue heron and eagle over my head, the *Mississippi Queen*, or the wind pushing the canoe down the main channel will flash before my eyes.

But I think my most lasting impression will be my first one, getting on the river in a canoe! I remember experiencing the absolute beauty of the river surrounded by walls of green tree-lined bluffs and blue sky. But most of all, I'll remember feeling the power of this majestic body of water . . . truly been a lifeline of this country with its power in its beauty.

We woke to the sound of Tom Ball singing "The Red Red Robin Goes Bob Bob Bobbing Along" at the top of his lungs. By the time he had finished, the ten of us had crawled out of our tents. I gazed out at the Mississippi River, not thirty feet from where we had pitched our tents. The river had disappeared into fog; looking at it was like looking off the edge of a high cliff into empty fog. The others began to strike their tents and load them into our two twenty-six-foot canoes. I then decided to wash off with a quick dip in the river. When I had finished wading, I helped Darrel and Jason load the food supplies into the canoes.

I cannot remember if I have ever been in mist as thick as it was that morning . . . . A few minutes later, we stopped on a channel marker to regroup. A channel marker is a twelve-foot, concrete tower that tells barge pilots where the boundary of safe water is. I climbed to the top of it, and from there, I could barely see the far bank. Even from there, the only life I could see was a great blue heron.

Once we reached the lock the river was no longer ours. At least ten boats joined us as we waited for the water to drain off. The air had cleared some and the river was simply normal again.

One approaches six days of canoeing on the Mississippi River with high expectations, and now, at some distance in time from the trip, those expectations still seem more than fulfilled. It is no wonder people write books about such trips. There are so many things one observes and learns and so many unique experiences. You want everyone to know what you have done and that you are a changed person. In spite of all the things one learns on the river, there seems to be more "what" and "why" questions about it than before.

Right from the start, one is surprised to learn the Mississippi River is not in a natural state but is largely a product of man's intervention. The lock-and-dam system is responsible for the location of the channel, the islands, and the wildflower-covered backwaters. Then one learns that even the river's current state is in jeopardy from invading soil, plants, and even small mussels.

In spite of man's efforts, there is an incredible amount of nature to be communed with along and in the river. The early, foggy mornings in the backwaters left the deepest impressions. The herons and egrets sitting sedately on dead tree limbs, the water covered with water lilies and lotus blossoms, and glimpses of a pair of foxes hunting along the shore are just the start of a long list of natural wonders sighted.

A very big part of this adventure was the human interaction, both within the group and with others encountered along the way. A group seems to have an intricate personality. This one developed many facets as it dealt with activity planning, route choices, work sharing, and even the personality quirks of its members. The variety of knowledge and talents of the group made the trip an enriching experience. Contact with people along the river who used it for their livelihood, avocation, or recreation kept the river from being an isolated physical object.

In the end, there seemed to be so many unanswered questions. Will the river survive our trying to control, use, enjoy, preserve, love, and develop it? What were all those unidentifiable birds? What is the biology of that yellow flower that only grew along the bank where streams emptied into the river? Can humans learn to live and work together in harmony? Why were all those large, dead mussels floating down the river? What can we ever do that will top an experience like this?

I have kept all these writings for almost thirty years and hadn't read them in nearly that long. If you live long enough for reflection, you'll find you did some things along the way that you're right proud of.

### Author's note

All of the above writings were written contemporaneously on the river. They were freely given to me by the paddlers themselves and never intended for publication. I have not seen any of these folks in many years and cannot remember who wrote what. I do find the writings to be a powerful testimony of what those folks felt a week-long canoe trip on the Mississippi meant to them.

## CHAPTER 15

# Sharing a Passion for the River

She did some politickin' that was tricky and hard
And got the pilothouse for the schoolhouse yard
—John Hartford, "Miss Ferris"

I was fortunate to have been touched by Ruth Ferris's boundless enthusiasm for the Mississippi River, though physical impairments had, by her eighty-seventh year, slowed her considerably. I remember the encouragement she offered me in my project—the simple act of uniting people with the river. Ruth Ferris spent most of her life doing the same thing, piquing a curiosity within her students, nurturing a desire to know more about the importance of the great, brown god, and gaining a knowledge of its history and a commitment to its present and future.

It cannot be said that Miss Ferris's love of the river was cultivated in a youth spent by a river home. Rather, she was born in 1897 in the flat cornfields of northern Missouri and gave the river little notice until she had graduated from the University of Missouri in the 1920s and commenced her career as a teacher in St. Louis. Always a student of folklore and primary sources, she had lucked upon her grandmother's written remembrances of a pre–Civil War trip on the Missouri River steamer *David Tatum*. Thirsting for more information on the boat, she was directed to Captain George W. Vaughn, a venerable Missouri River pilot. Knowing what I know about old pilots and young ears, I can only speculate about the tales that must have been told at the Masonic Nursing Home in those pretelevision days. It was surely the genesis of a love for the river and its lore that was to make her the most highly regarded "riverlorian" of her time.

Like so many who have experienced the subtle, unyielding grip of the river, she soon found herself on a one-way trip, following the water's muddy call. Her summers were spent travelling the waterways on the few excursion steamboats that still graced the St. Louis riverfront in the 1930s. The late Jim Swift, former business manager of the *Waterways Journal*, who wrote its Old Boat Column, fondly recalled the many memories he shared with Miss Ferris. "She was the unofficial leader of our group that spent long summer days riding boats, up to St. Paul, down to Memphis and Chattanooga. Her enthusiasm never once waned for the river." She seemed comfortable just walking into the pilothouse and starting up a conversation with the pilot or captain, like they were old friends, and they usually were. Swift also remembered their trip aboard the *Golden Eagle* up to the Twin Cities, the first passenger-boat excursion through the Mississippi's new system of locks and dams.

Miss Ferris's attraction to the river did not end with the clanging of the first school bells in September. Anxious to share the river with her fifth-grade students at the Community School in St. Louis, her experiences of the summer became the pine tar and coal that fired the boilers in the dreams of her charges. The other faculty members at the school soon learned that whenever they saw an older, stately gentleman approach the school that it must be another river captain coming to speak to Miss Ferris's class. She also christened the Hot Stove Navigation League of America for those whose fix could not wait through the long months of winter for the opening of the river. The St. Louis chapter was known as Scuttle #1, and its insignia was a potbellied stove.

Perhaps Miss Ferris's best-known student was the late singer-songwriter John Hartford, who credited her with bringing the river's world to life for him, a passion that he never lost. In his song, simply entitled "Miss Ferris," Hartford recounts his teacher's most audacious journey upon the waters of river history. Miss Ferris, distressed but undaunted by the sinking of her beloved *Golden Eagle* at Grand Tower Chute in 1947, "did some politickin' that was tricky and hard / and got the pilothouse for the schoolhouse yard." Salvaged from the Mississippi's swirling silt, the wheelhouse of Captain Buck Leyhe's erstwhile steamer began its new life as a living artifact from the river's past. It sits today in the Missouri Historical Society in St. Louis.

Miss Ferris retired from the Community School in 1957 to spend more time pursuing her river passions. She was curator for river museums for the Missouri Historical Society and aboard the steamer *Becky Thatcher*, which was located at the St. Louis riverfront for several years. During this stage of her life, Miss Ferris's reputation as a historian and lecturer blossomed, and she became the acknowledged authority on the subject.

John Hoover, director of the Mercantile Library, recalls the nurturing hand of Miss Ferris when he was curator of the Herman T. Pott Waterways Library, which is part of the larger library: "She bought me my own personal copy of Way's Packet Directory. It was my rite of passage into the world of the river, and she was proud and beaming to be able to do it." Miss Ferris's reputation as a riverlorian was not confined to St. Louis. Hoover remembers spending an afternoon with her aboard the *Delta Queen*: "She marched on the *DQ* like she owned it, and gave me a tour of the boat. We spent hours going over the boat and river history."

Lest we imagine, however, that Ruth Ferris's love for the river left her one dimensional, Hoover told me that, on a typical day in her living room, one might find a *National Geographic*, a novel, and a textbook on ancient Greece, alongside her latest copy of the *Waterways Journal*. She felt that fame and money were transient but that true power resided in knowledge. She saved everything because she considered objects to be the physical manifestation of history. She became lifelong friends with river historian Keith Norrington of New Albany, Indiana, who wrote, "She found myriads of ephemeral items that most would have overlooked. Each one had its own story, whether it was a dusty book found in 1930 or a plaster cast of her own foot sunk into the Mississippi's mud."

Upon retirement from public life in 1970, Miss Ferris moved all her possessions to her home, where she held court from her basement, which she termed her "E. Z. Rocker." In the 1980s, she began moving her collection, so it could be shared for the future. I had the opportunity to peruse it, time to peck only the outer skin of her collection. There was gingerbread from an old steamer, a model of a towboat and its barges, and a scrapbook of flood pictures—not the flood of 1993, but the torrent of 1937. Even an incomplete tour of its vast contents would take weeks.

Ruth Ferris was the quintessential teacher—in love with life and learning. Like Mark Twain's legendary death (and birth) in the year of Halley's Comet, Ruth Ferris died at an epic time, in July 1993, during the greatest Upper Mississippi River flood of all time. Though she can no longer pass her days with friends in the E. Z. Rocker, her memory is always there in her love of the Mississippi and its people.

While most of the City of St. Louis seems to resolutely turn its back on the Mississippi River, the Pott Library beckons to those with imagination enough to visualize the ghosts of voyageurs, flat-boat men, keel-boat men, and steamboat pilots walking along the downtown streets. Inside the library's silent catacombs, hidden within a labyrinth of rare manuscripts telling the story of the city, resides the living collection of a remarkable woman. However, the thousands of letters, files, newspaper articles, postcards, models, and artifacts that Ruth Ferris bequeathed the library represent only the physical manifestation of her legacy. Her true spirit resides in the memories of the many people who knew her as a friend.

### Author's note

I wrote this article for *Big River Magazine* about twenty-five years ago. The St. Louis Mercantile Library and the Pott Library are now located on the campus of the University of Missouri–St. Louis—and not downtown.

# CHAPTER 16

# Of Time and the River

As a child, I remember going with my father to Hall Street in north St. Louis, across the railroad tracks, through areas of broken glass and debris, and down to the river. My dad would sit on a usably comfortable hunk of riprap, drink his beer, and watch me hurl rocks into the café-au-lait-colored water. The river dwarfed me with majesty and gripped me with mystery.

Years later, I found work as a deckhand on a towboat. My job was to mop and sweep decks and to carry heavy things on my back and shoulders. It was dirty, sweaty work and dangerous at that, slipping on icy winter decks, risking frostbite and hypothermia. In the summer, the same steel barges would burn holes in my jeans, and even several fruit jars of lemonade and ice water barely kept me hydrated in the relentless, searing sun. Not often, but there were times that our tow ran aground or swiped a railroad bridge. Barges that, seconds before, were safely confined by steel rigging, suddenly had minds of their own and sought to take separate paths. To a young, single man searching for adventure, towboats were the perfect place.

For eleven years, I worked as a deckhand, mate, and pilot. Some of the work I dreaded. At times, I was engulfed by isolation and loneliness. However, in total, I had found a place, a way of life that was my own.

I found something else in those years, too. I discovered the beauty of the Upper Mississippi River. Its islands, sloughs, and storybook towns were like a dream. Here, we were adventurers on Sinbad's eighth voyage, and I wished that those trips would never end, that we would keep going past Dubuque, LaCrosse, Red Wing, past the locks and dams at Minneapolis to where the waters of this river begin.

One summer, I took my canoe and lived on the Upper Mississippi, exploring its backwaters, catching its panfish, watching its sunsets, sharing coffee and Leinenkugels with its residents in places like the Trempealeau Hotel and the Dam View in Alma, Wisconsin.

I want the river to have clean water. I want those fish to still be there and worth catching. I want those islands and sloughs to still be places for heron, egret, pelican, and eagle. I am concerned with my environment. I drink the water of the Mississippi every day of my life. And I still want there to be towboats! I do not think that towboats and a healthy environment have to be mutually exclusive. Towboating is part of my heritage, part of who I am. But above that, it is part of what this country is, too. People were using water to transport themselves and cargo before any of our ancestors (excluding Native Americans) came to this political entity we call the United States.

People I have ridden towboats with, sharing adventures, hard work, rich and delicious meals, and marvelous 3 a.m. stories are a part of the treasures I know as life. I named my son after a towboat captain. I would never support any program that would call for the elimination of towboats on the river.

Perhaps I am overly optimistic, but I see a river in which towboaters and environmentalists can work together. As a matter of fact, it has already been done. At one time, towboats did discharge raw sewage, bilge oil, and trash into the river. These past sins are undeniable but not unprecedented. Steamboats, log rafts, pioneers, and even American Indians committed some of the same acts. Modern towboats, run by responsible carriers, hold their sewage in tanks, burn trash in incinerators or put it shoreside, and do not pump oil into the river. Cases of environmental damage are few.

There have been numerous studies done concerning costs to taxpayers and the environment, labeling river maintenance as a "subsidy" to the towing industry. I am not a political expert and do not have the time nor inclination to read studies in their entirety. However, I know that a fifteen-barge tow of coal or grain carries the equivalent tonnage of between seven hundred and eight hundred tractor trailers. Anyone would be hard-pressed to convince me that I would be better off with eight hundred more trucks on the highway.

This is simply one man's opinion based on a life of living with the river, accepting it on its own terms. I love it, I want others to love it, and I want it to always be there and be clean. I think we can have a river that supports wildlife, fishermen, recreational boaters—and towboats. The river can give its gifts to all of us.

### Author's note

I wrote this in the early 1990s as a letter to the editor for the St. Louis Post Dispatch. Surprisingly, they put it on the editorial page.

# CHAPTER 17

# Reflections on the 1993 Flood

In 1975, I was the first mate on the towboat *Susan B*. I came on one midnight watch with the towboat and our fifteen barges tied off in the fog a few miles above Lansing, Iowa, on the Upper Mississippi. The previous watch had used our small, motorless jon boat to reach the bank to find a suitable tree to tie off to. Instead of returning the boat to its secure place on the *Susan B*., they had left it on the stern of the tow.

Having just awakened and poured down two or three cups of coal-black coffee, I was full of spit and vinegar and ready to take the jon boat back to its home. I was determined to get that boat back on my own skill and power. We had two long "wing wires" that ran from the outside corner of the tow back to the towboat, and it seemed a relatively straightforward task to stand in the jon boat and pull it back along the wing wires to the *Susan B*. After all, I had seen Norris Frederick do it. But he was from southern Louisiana—and I wasn't.

The current seemed to be running lazily along and did not appear as though it would offer much resistance to my plan of action. Things went pretty well for the first ten or fifteen feet as

I pulled myself hand over hand toward my destination. Soon, though, my arms began to lose some of their strength, and my watery path turned the jon boat more sideways to the current. My forward speed stopped, and the boat began to lean toward the force of the water. The more I tried to exert my body, the more the boat continued its listing action. Soon, a few trickles of greenish water began to come over the gunwales of the boat. I did not immediately realize it, but at that point, the story was already told—nothing was left but to act it out.

Events transpired so deliberately that there seemed to be ample time to save myself and the jon boat. I just knew that I could surely straighten the boat out if I could only get one second when that current would stop hitting me. But it never gave me that second because it is not the river's nature to stop. Unlike waves or powerful gusts of wind that offer peaks and valleys of energy to catch your breath and prepare for the next onslaught, the river is constant and relentless. I could see the scene and its climax clearly laid out in front of me, and there was not a damn thing I could do about it.

First, my knees were wet, then my thighs. Soon, the river was completely around me, inviting me into its arms, demanding that I become part of it. Luckily for me, I still held the wing wires and was able to pull myself to within arm's length of my deck mates, who lifted me to safety. I watched with empty embarrassment as the jon boat flopped under the barges, filled with water, and headed down through the mist toward Lansing. We would never see it again, and I was sure the captain, Turnip Hartnett, would fire me as soon as he got up at 6 a.m. He didn't like to spend any more of the company's money than he had to.

It was a long night as I lived with the teasing of the crew and tried to figure out what my first words would be to explain to Turnip what had happened. I shrewdly gave him time to finish his first cup of coffee, then told him the story. He sipped calmly, then asked, "Did you get wet?"

"Yes," I said.

"Good." A few moments of silence ensued. "Did you learn anything?" he continued.

"I-I s-suppose so."

"Well, why don't you go to bed and see if you can't figure out what you learned."

I was so amazed that he hadn't told me to pack my bags that I was much more in a relieved than reflective mood. Later that morning, I told him something awfully inane such as, "I learned never to do that again." Luckily, he let the whole thing drop. We had another jon boat, and it may have been too much trouble to get a replacement mate to Lansing.

Eighteen years later, I was much weaker, hopefully a bit smarter, and definitely more aware of my own limitations. I still worked on the Mississippi, though on a floating casino in Davenport, Iowa.

About the middle of June, we heard news reports of ten- and twelve-inch rains occurring in Minnesota and Wisconsin. We saw pictures of flooding on rivers such as the Black, the Root, and the Zumbro. Intellectually, we recognized that this water, springing from every pore of the saturated earth, would eventually make it to Davenport. However, there is something inside a person that will not admit to a fact completely until he sees it. As we prepared ourselves physically for this wall of water, mentally and emotionally, we were not ready for it, even after it got to us.

The river slowly began to rise a tenth or two tenths of a foot each day. Then the volume increased to half a foot. Soon, we could not get the casino boat under some of the bridges. Our parking lot flooded, forcing us to move to a different landing. We could no longer cruise on the river, but we could still take passengers in the casino, which was the focus of the business anyway. At this new landing, we somehow assumed we would be safe because, after all, even the record flood of 1965 had not gone over the floodwall there. But the skies continued to open up every day . . . and that water just kept coming. Relentlessly. Our smug rationale was flawed—nature does not keep records, and records are made to be broken.

The river crept higher up on the wall, seemingly by the hour. It soon became apparent that the unthinkable was about to occur. Logically, perhaps, there was a certain festivity to all this in the local media. Every day the newspaper would have a full-page headline highlighting numbers of the predicted crest, which changed daily . . . upward. Some enterprising entrepreneurs began selling T-shirts about the great flood. President Clinton arrived one Sunday evening and, with his motorcade, stopped at points along the river with Agriculture Secretary Mike Espy.

Soon, we were forced to move the casino again. However, this time, it was to a landing where passengers, even gamblers, could not be boarded. The opulent casino with its seven hundred slot machines, its roulette, craps, and blackjack tables, its thick carpeting and mirrored bars with video poker games sat silently and eerily alone with only a skeleton security and deck force aboard. Six hundred and fifty people were laid off. Being in the marine operations department, I was one of the lucky ones to keep working. Many businesses in town suffered as badly or worse than us. Fittingly, the record crest came late and was greeted not as part of a big bash but, as it truly was, a terrible hangover. And still that water kept coming . . . relentlessly.

Then, one night, comfortably watching television in my hotel room, I saw something that drove home to me the fact that devastation is a relative concept. A CNN reporter was interviewing a young boy from Hannibal. They were in a motorboat, and they were going past the boy's home—or what used to be his home. The boy's family had seen those same reports of rain up the river, had prepared sandbags, had done everything humanly possible to save their home. And they, too, were not ready for the reality of what was to come. There was a shock on the boy's face yet a certain sense of toughness, of knowing that they did all they could do and would deal with what was to come next . . . together.

This boy, just a few years older than my son, stared blankly at his home, his bedroom full of baseball gloves, model airplanes, and games, his garage with his bicycle, his kitchen where his mother had cooked hot biscuits and gravy for him, and his warm living room with their television. All were gone. And yet the river kept coming . . . relentlessly. Maybe if it had given them just a short break, they could have shored things up, saved that house full of all their possessions, and their memories . . . maybe they could have.

The river had begun to fall in Davenport, just as it did later in Keithsburg, Burlington, Keokuk, Hannibal, and St. Louis. Then the real work of cleaning up began. Our casino went back in operation and some passengers, when they took their minds off the tables for a moment, could look out over the still-swollen river and reflect momentarily about where all that water came from and where it next went. I prayed for that eight-year-old boy in Hannibal, that he was somewhere safe and dry. And I thought about that jon boat lesson Turnip Hartnett was trying to make me understand in 1975.

The year of 1993 still is the record flood on most of the Upper Mississippi River.

# CHAPTER 18

# Just Havin' a License Don't Make a Man No Pilot

Once upon a time, in the ancient days before electronic charts . . .

Two old boys sat blandly in the pilothouse of the *American Queen* on a sultry Mississippi afternoon.

"Captain Lee, did I ever tell you about how Neckbone got turned around one night an' didn't know it?"

Got a little story to tell, from the Corps of Engineers days, and it goes like this. Captain John Dugger, being a recent corps retiree, had an endless supply of them.

"Neckbone" was one of his more reliable protagonists. Neckbone was an undaunted adventurer and quite resourceful. In charge of ordering supplies and food for over five hundred crew members, he never spent much of his workday in the pilothouse of any of the six attendant boats learning the secrets of navigation. But he had gotten sufficient "sea-time" letters to "sit" for a towboat operator's license. Being "book smart," he passed the Coast Guard exam and, because companies were desperate for anyone with a license in the 1970s, was soon piloting a towboat on the Lower Mississippi River.

He was so lost in the practical part of this endeavor, though, that he had to have his paper chart in front of him the entire time he was piloting. During daylight, "road mapping" was not much more than a minor inconvenience. At night, however, when you had to constantly turn your flashlight on and transpose the page from the bulky, clumsily bound chart onto the veiled darkness in front of you, it was an awkward practice at best.

Well, Neckbone only had a single, empty barge in tow one particular night but had to go a long distance down the river with it. There was another large tow ahead of him that he had to overtake, which he did without incident. Then disaster hit. His chart and coffee cup fell on the pilothouse floor, and he had to reach down to clean the mess and gather the coffee-soaked chart back up. In that brief period of distraction, his boat and small tow got turned around and was headed back up the river. Unaware of what had happened, he returned his attention into the lonesome night. He was suddenly faced with the navigation lights of what he thought was an upbound tow! He anxiously called the boat to make a passing arrangement.

A pregnant pause ensued.

A laconic voice on the radio finally pierced the shadows of the pilothouse. "Man, ain't you the one just got around me?"

Adjusting back to situational awareness while still breaking out in hives, Neckbone made a shrewd command decision. He would be best served by turning back around, getting back behind his new-found buddy, and just "bird dog" (follow) him for the rest of the night.

As many of my older pilot mentors have told me over the years, "Son, jist havin' a set o' license don't make a man no pilot!"

But it is most certainly needed to legally pilot a commercial boat on the river. And the process of getting one is in a nearly constant state of revision.

Dave Calvert, head of the Regional Exam Center (REC) in Memphis, informed me that the St. Louis REC still retains a copy of Mark Twain's original pilot license. My curiosity suitably stoked, I googled it and found a copy of a license issued to Samuel Clemens in 1859 in accordance with the Steamboat Act passed by Congress in 1852. After 1811 (when the steamer *New Orleans* made its historic voyage) and until 1852, there was little regulation over steamboats, with a "have-at-it" mentality prevailing. This led to many collisions, explosions, and other tragic occurrences.

The towboat industry followed this same absence of documentation until 1972 (coincidentally my first year on the river), when the Towing Vessel Licensing Act was passed. In an ironic twist, pilots on steam towboats had previously been required to have an extensive license, while those on diesel towing vessels were exempt from any regulation.

I, like Neckbone, was in the right place at the right time and seized the day with my first license in 1975. I went to a training school in Memphis and quickly learned that you had better give the instructor correct answers because he was a deft strike thrower with a piece of chalk aimed at your head. I passed my twenty-five-question, multiple-choice test and received what was known as an "operator of uninspected towing vessel" (OUTV) license. A straight-vanilla license, it was equally valid for Neckbone's one empty barge or for forty barges on the 10,500-horsepower *Cooperative Spirit* and on any stretch of the western rivers (the Mississippi and its tributaries).

When Sam Clemens got his license from St. Louis to New Orleans, he was issued paper for that stretch of river and only that stretch of river. The steamboat inspectors from the St. Louis District would ask him questions pertinent to that specific route until they were satisfied that he could safely get a vessel up and down his area of expertise. It should be noted that even well-honed pilots in those days could meet with tragedy as the river was constantly changing and steamboats, with wooden hulls and fiery boilers in close proximity to each other, were inherently fraught with peril.

I have often heard folks say they know someone with a "pilot's license." In truth, few possess a first-class pilot's license. Only by intimate knowledge of a section of river can that license be obtained. It is left over from the steamboat days and is now only required if one is aboard certain larger vessels in "pilotage waters." Licenses are currently issued by the United States Coast Guard, and an applicant for a pilot's license is required (after a requisite number of roundtrips) to "draw the river" on a blank piece of paper. The applicant must list navigation lights and their characteristics, bridges (with horizontal and vertical clearances), sandbars, dikes, wing dams, revetments, power-line crossings, submarine-pipeline crossings, and other pertinent information.

Fortunately, God blessed me with a quaint knack for remembering numbers. In the early 1990s, I drew over 2,300 miles of river, basically St. Paul to New Orleans and Cincinnati to Pittsburgh (Ohio River), from memory. It took me over a year, but that was the license I needed, at the time, to pilot the *American*, *Mississippi*, and *Delta Queens*. Such a license is now only required from Baton Rouge to the Gulf of Mexico on the Mississippi River, making most of the first-class-pilot section of my imprimatur somewhat like a Belgian lampshade in my wife's Victorian living room.

Presently, on the river, there are two major subsets of license: (1) towboat and (2) inspected vessels (usually passenger). Replacing the OUTV, there are now master of towing vessels (MTV), mate/pilot towing vessels, and a steersman's license for trainees. Like the erstwhile OUTV, these licenses are valid for any size of towboat. Every towboat must have at least one MTV holder on board, and a steersman must have a higher-licensed individual present with them while at the controls. These exams are no longer twenty-five questions but have five different modules (Rules of the Road, Navigation General, Plotting, Deck Safety, and Deck General). On the bright side, due to sensitivity issues, the instructor no longer throws chalk at you for wrong answers.

The licenses for inspected vessels (this nomenclature is no longer accurate since towboats are now inspected, though at a different level) are issued in gradations from twenty-five tons up to any gross tons (unlimited). Don't be like this writer was and think vessels are put on a scale to be weighed. Tonnage does not indicate weight. Rather, it is a volumetric calculation. This sometimes results in certain mariners being "pencil whipped." Although it is an expensive process, an owner can take a boat registered at 2,000 GT (requiring a higher license), make a few internal structural changes, and voila, you have a vessel of less than one hundred GT. Because there are more people with a smaller license, the supply-demand curve has swung to the owner's behalf, and he can pay folks less, not to mention less in inspection and documentation fees.

There is also an operator uninspected passenger vessel (OUPV) license. It is colloquially known as a "six-pack" license and has nothing to do with beer or abdominal exercises. With this license, an operator can take out up to six paying guests on a vessel not subject to Coast Guard inspections.

Other than the fee for training school, my first license cost me nothing. In an attempt to raise funds, however, the Coast Guard now charges for license issuance, up to hundreds of dollars for original licenses.

Certain other changes have been made in licensing over the years. Subsequent to the Big Bayou Canot (Sunset Limited) train accident in 1993, mariners were required to pass a "radar course." Though the strategy may have been sound, this tactical, knee-jerk implementation created a course that was costly and of little practical value to most mariners. It was opportune for training schools but has since been thankfully altered in its scope.

Finally, on behalf of professional mariners, I have to ask, "Why is the recreational craft owner exempt from licensing requirements?" Some of them seem to drink alcohol in abundance (without much fear of a DUI) and destroy riverbanks with their wake (no worries there about a speeding ticket). They are capable of causing death and mayhem yet are impervious to regulation and don't even have to pass that twenty-five-question, multiple-choice test, like Neckbone had to.

# CHAPTER 19

# An Unscheduled Pit Stop

I feel lucky about having two spirits I would consider to be my best friends. One of them, my wife, Dianne, is unfortunately not always in agreement with me on certain things. The other, my black Lab, Charlie, as long as the treats keep coming, voices no complaints at all.

I am of the opinion that when God put dogs on the earth, he created the almost perfect creature. I say "almost perfect" because of one offensive ancillary detail. If I lived out in the woods away from all people, this detail would not even present a problem. But why can't their poop just evaporate or dehydrate and blow away? In Minnesota, in January, fulfilling one's civic duty of carrying plastic bags around with you can become much more than just an annoying nuisance. Even emperors and other royalty throughout history never received such scatological subservience.

Well, you may ask, why is this in a book about river boats? Before I go further, for other dog lovers, rest easy; you can keep reading. It does not have a sad ending as many dog stories do.

In 1995, I had completed the agonizing effort of drawing over 2,300 miles of river, and I could go to work for the Delta Queen Steamboat Company. I was in line to actually earn a paycheck to give Dianne some return on her investment. It should be noted that, at that time, this affable raconteur was "only" about forty-five years old. In the view of the other steamboat pilots with the company, many of whom could recall the 1937 flood and World War II as something besides interesting historical perspective, I was a mere child. With no gray hair sprinkled about my head, my bona fides was eyed with a great deal of skepticism.

I was living in St. Louis then, and my first sho'-'nuff piloting gig with them was to catch the *Mississippi Queen* down by the Arch and take it up to Hannibal and back for a brief three-day cruise. Captain Buddy Muirhead must have been primed and prepped by the old guard pilots as, after an introductory handshake, he asked to see my license, which he inspected with great care.

My first night aboard the *MQ* was uneventful as we cruised up by Alton, Grafton, then toward Lock 25. I had never handled a steamboat before in my life, so I was a bit uneasy at first, worried I might actually have to do something out of the ordinary, like stop. Luckily, nothing like that happened, and I went off watch near Kinder's Restaurant by the Golden Eagle Ferry. I had stood my first noneventful watch on a steamboat!

The front watch made Lock 25, and I came on looking at Clarksville and 24 the next afternoon. After successfully navigating the lock, under the expert direction of Captain Muirhead, then the Louisiana, Missouri, railroad bridge, I was getting some notches on my belt and feeling pretty darn good about this whole steamboatin' thang.

It was a perfect day: warm, sunny, light breeze, almost all the passengers sitting out on the deck enjoying the early summer afternoon and verdant landscape. All was good with the world. Then Captain Muirhead came into the pilothouse. He had a perplexed look on his face.

He called me Captain Hendrix, and I was feeling suddenly mighty empowered. He had never run the Upper Mississippi very much, and he was asking me for advice!

"Captain," he said, "can we get into the bank up here anywhere? Like, maybe, at that island up ahead?"

I looked up at North Fritz Island and remembered that, in my first green years of piloting tows, I had almost been set up into the trees there with fifteen coal loads. It was only by dumb luck and the grace of God that I had kept from crashing. Other than by accident, I couldn't imagine why anyone would go in there, but yes, I did have some first-hand experience. However, I didn't share all the neurotic details of that night with him. I didn't want to scare him off again.

"Yes, Captain," I said. "We have ample water at this stage. I don't think it would be a problem." Then I waited.

"Captain, thank you," he continued. "We have a lady on board who brought her service dog, Brody, with her. The dog is very well trained, so well trained that he won't go to the bathroom on board even though she brought a box along for him to use. The poor dog is starting to look a bit stoved up, and it's going to be several hours before we get up to Hannibal. I feel like I should put him off up here to take care of his business and get him out of his discomfort."

Well, having explored many of those same islands by canoe, I racked my brain to remember if I had ever gotten out to pee at that particular one. "Captain," I said, after some thought, "I think it can be done."

Captain Muirhead called down to the engineers and told them to be ready for some bells, then he went out to the bridge wing to give me orders for the engine-order telegraph. After a few bells, we had successfully maneuvered the *MQ* into the bank. I left the pilothouse to come out on the bridge wing with him for the show.

My recollection is that the *MQ* passenger capacity was about four hundred, and every damn one of them seemed to be out on deck to see what was going on, staring quizzically up at Captain Muirhead—and me. I could almost imagine the murmurs circulating about. "Why are we stopping?" "Must be engine problems," an instant expert would speculate. Or "I bet that kid pilot ran aground." "Maybe he got drunk—or stoned!" "Am I going to miss my tour in Hannibal?"

The moment of truth was at hand as the stage was lowered down into the brush and weeds on the bank. Soon, a crew member could be seen ceremoniously walking Brody down the stage and onto the bank.

Captain Muirhead turned to me and said with an air of chagrin, "If that dog don't [poop] soon, I'm afraid I'll have to shoot it." I smiled back and chuckled uncertainly, checking him for a firearm. I had heard he was a man of the cloth, so I was sure he was speaking to me in jest. The dog, as they will do, sniffed and scratched around interminably, and the tension grew.

Expressions of puzzling concern by the passengers turned to laughter as the gallant hound finally squatted down, dug in, and relieved himself on the bank. It gave new meaning to the term "poop deck" for me as Brody also whizzed on a few trees, sending out "pee-mails" to any other canines in the area. I find it hard to believe that any dog, before or since, has had a more sympathetic audience for its healthy eliminations. He was brought back aboard the MQ to a thunderous round of applause. I don't recall anyone being worried about having plastic bags along either.

At the captain's dinner, Captain Muirhead made a special Milk-Bone toast to the gallant Brody.

# CHAPTER 20

# Maneuvering an Elephant in a Bathtub

In the summer of 1996, Captain Dick Karnath and I were the first pilots to take the *American Queen* up the Mississippi to St. Paul. On the trip up from St. Louis at most of the locks we transited, we were greeted by throngs of curious and enthusiastic spectators. I assumed they were enthralled by the size and splendor of the *AQ* (420 feet long and ninety-five feet wide), but at times, the notoriety made me uneasy. Fame can be a fleeting friend.

Some of the company's older pilots had predicted disaster from taking the huge boat on the narrow Upper Mississippi with pilots too green for the task. I did have some concerns about the former but not at all about the latter. Dick and I were both well out of yellow diapers, and thanks to Z-drives and bow thrusters, the boat handled well enough to make pilots look good, even if they weren't.

We arrived in downtown St. Paul on a sticky morning, a bit late due to the weather. Norm Coleman, the mayor of St. Paul at the time, was waiting with his entourage at Lambert's Landing to officially welcome us. We normally landed with the bow pointed downstream at St. Paul to allow for a more expeditious departure later in the day, but this was the first time we had to turn the *AQ*'s massive length around in such a confined space. In fact, there was less than one hundred feet of wiggle room, with a concrete wall on one side and fleets of barges on the other.

The barge-fleeting company helped us by narrowing up their strings of barges, which at least made the maneuver possible. However, we had to turn without being able to see the bow or stern of our boat and how close we were to either side. This necessitated putting a trusted crew member with a walkie-talkie on both ends of the *Queen* to let us know how close we were to the concrete wall on the stern or the barges on the head.

Captain Adrian Hargrove was on watch with me, and in his west-Kentucky drawl, he suggested, "Son, let me turn it around. I've done it more times here, and the mayor looks like he's getting hot and sweaty out there."

Lesser men have put up much better fights than I did that morning. I quickly grabbed a walkie-talkie and headed for the stern, without giving him time to change his mind. With his usual grace and aplomb, Adrian gently put the boat into the landing and got the mayor out of the weather.

We had made it up to St. Paul unscathed, but now faced the challenge of bringing the behemoth down the river—a trickier feat. I was on watch when we departed and steamed down toward Pig's Eye Bridge. Dayton's Bluff, overlooking that stretch of river, was a multihued mass of parked cars and humanity, there to watch us—me and the *Queen*. The bridges ahead were too low to pass under without hitting the pilothouse, so I was standing outside piloting the boat from the bridge wing—a set of controls on each side of the top deck that we use when the pilothouse has to be lowered into the deck to clear low bridges and electrical lines.

John Davitt was one of the company's smoothest and most trusted captains. When he strode purposefully out to the bridge wing, I figured he was going to provide me with some tactical wisdom concerning that narrow little railroad swing bridge we were headed toward. He did, sort of.

"Don't hit it," he advised. "You've got too many witnesses."

With this priceless bit of consultation firmly wrapped around my brain, I wondered if anyone on the bluff had binoculars powerful enough to identify me if I planted the boat around a bridge pier or put the bridge turntable out of commission for a few weeks.

Thankfully, I didn't generate any breaking news that night. However, it launched us into a series of adventures of the sort that we encountered taking the *AQ* anywhere other than the Lower Mississippi River.

The stacks on the *AQ*, at that time, were about 115 feet above the water level. Even with the pilothouse sunk into its cubbyhole beneath the upper deck and all the other accoutrements—such as stacks, radar, radio antennae, etc.—pulled down, we couldn't get much lower than fifty-five feet above water level—fifty-seven or so if you counted our heads and shoulders while we stood at the bridge wings.

The unimaginative builders of most bridges on the Upper Mississippi, Ohio, Tennessee, Illinois, and Cumberland Rivers had failed to foresee our existence and recklessly built their projects with clearances of sixty to sixty-five feet in "pool water."

At this point, I should explain the term "pool water." It is, at best, a subjective metric used to anticipate the vertical clearance between the water and the "low steel" of a particular bridge. However, pool water can mean different things at different parts in a pool. For example, when the river is rapidly rising and dam gates are completely opened to flush the excess out of the system, the area right below a dam can be several feet or more above "pool," while right above the dam, water can be a few feet below "pool."

On a vessel as tall as the *AQ*, the mathematical gymnastics for bridges and power lines are never ending. Often, we just had to ease up or down on a bridge and eyeball it to see if we could clear. I remember Captain Karnath and I exposed out on the bridge wing coming down on the Hastings railroad lift bridge one dark-ass night—maybe a bit too fast—thinking we had several feet of clearance. The meaning of "low steel" is a vivid concept when you are suddenly at eye level with that steel and your toes start to curl as you realize there is no immediate escape other than to scream "*Duck!*" and "do the ostrich" to avoid decapitation.

Probably one of us had calculated incorrectly, and it must have been me since he used to teach math. I prefer to blame it on the bridge tender, who obviously didn't raise the trestle quite high enough.

On another occasion, I was not aware that they were painting the Prairie du Chien, Wisconsin Bridge—a particularly low one. We approached it at night, so no workers were under the bridge, but their scaffolding sure was. Fortunately, I was going slowly enough that I could reverse engines down to the current's speed and just float through with my head in the full duck position while hanging onto the railing and nearly scraping my knuckles on the scaffolding planks just above me.

We were once trapped between low bridges in Cincinnati for over two weeks waiting for the river to fall. This cut deeply into company profits because tours had to be canceled and monetary adjustments made to disappointed passengers.

A year or so later during a rainy period on the Ohio, with that event still fresh in mind, we sat in Maysville, Kentucky, upriver of Cincinnati, waiting for shore tours to return. We were moored between two bridges: a newer one about two miles downstream and an older one just upstream of our landing. I watched the river rising on the ice breakers that sat below the old bridge. All day it poured like "piss from a boot," with dire predictions of tornadoes later in the day.

When I came on watch that evening, tornado sirens were wailing, and the sky was green and yellow. I have never again seen a sky that pale, week-old-bruise color. I expected Auntie Em and Toto to come cartwheeling across the decks as I ducked into the pilothouse. Let the casualties be damned, we were heading to Cincinnati! Maybe.

The rain continued, and the only water-level data I had on the "Big Mac," Purple People, and Taylor-Southgate Bridges downriver at Cincinnati had been issued by the National Weather Service earlier that morning. Clearances on Ohio River bridges are more precise than the Upper Miss, but this data was old. So, how good was it? If it was good, we would have a couple feet of clearance. There are vertical clearances painted on some bridge piers, but at night, they are impossible to read until you are right on top of them.

As I approached Cincinnati, I decided that the safe course of action was to turn the AQ around and float backwards through those three bridges. My reasoning was that if I got down to the bridges and saw we wouldn't fit under one, I could shove back up the river and come up with a plan B. As an airplane pilot lands his craft into the wind, a river pilot has better control of a boat shoving into the current rather than backing against it. The last thing I wanted to risk on a fast-rising river was to "lose my stern" in between those bridges—the stuff nightmares are made of.

It was ten o'clock at night as the lights on the yellow arch of the "Big Mac" bridge came into view. There was a fashionable restaurant a quarter mile or so upstream of the bridge, and it being a Saturday night, it was busy. The rain had stopped, and men in polo shirts and women in black dresses and high heels were out on the deck making small talk and balancing their wine glasses provocatively between their fingers as our floating palace came down the river. Abruptly, the bow of the boat swung around toward them as I began the turn. I heard some screams and shouting. I wonder how many of them contemplated dropping their glasses and fleeing as our gangplanks approached them like jousting spears. "What manner of madman is driving this boat?"

As I completed the turn and pointed the bow of the AQ upriver, their fears turned to curiosity. Again, I was forced to abandon the comfort and warmth of the pilot house and navigate from the bridge wing, leaving myself in full sight of the crowd. I believe that they were even toasting the magnificent boat as it floated past them backwards. They were still shouting and hooting at us. I couldn't really make most of it out, but they were certainly attentive witnesses.

The data turned out to be fairly precise. We squeezed under the bridges and made it to our landing only to find the dumpster that had been left for us had been swept away by the river. Due to the rising water, we couldn't stay long enough for land tours and didn't want to get trapped between bridges again. So, like waterborne gypsies, we fled down to Louisville for more adventures.

# CHAPTER 21

# Eulogy to a Legend

As I drove back to St. Louis from Paducah, the sky began to cry. Initially, I interpreted the rain as a symbol of sadness. The heavens, fed from below by the river, were weeping at the loss of a great riverman. Soon, however, my thoughts began to change. He would not have taken it that way. "Son, look to the sunny side of life," he would have said. In reflection, then, I realized mine were not tears of sadness. It was angels shedding tears of joy that Adrian Hargrove would be joining them. That was the way he would have preferred to look at it, though his modesty would never have allowed him to flatter himself so.

He will speak to the angels now, like he spoke to the river for so many years. And the angels, like the river, will listen. It's not that I have witnessed the work of every great pilot that has ever been on the river, but he was the best that I ever knew. There may have been some that understood the principles and physics of leverage, motion, and current better than he did. I don't know of any, but there may have been some. And there may have been some that could quote mile markers and bridge clearances with more accuracy. Again, I don't know of any, and he had over 4,000 miles of pilotage endorsed on his license. But no one I ever saw danced with the river in his arms like Adrian Hargrove did. It was never a struggle for him to pilot a boat or a tow. Wherever he was, whatever vessel he graced, he was always waltzing with the river's flow—getting into its vibes, gliding with it, never fighting it.

Whatever meager expertise I shall ever attain as a navigator is mostly because of him and others like him. There were other great pilots who taught me things and taught me well. And Adrian put it all together for me. Like any fine artist, there was no wasted motion in anything that he did. It was simply being one with the river. He seemed to always be saying, believing "You take care of the river, and she'll take care of you."

Not that he never got in trouble. I remember once, over forty years ago, when I was learning to pilot on the Upper Mississippi, I stayed up off watch to observe his approach into Lock 19 at Keokuk, Iowa. Every air conditioner in Iowa and Missouri must have been running on high, the Union Electric Power Plant sucking all the water into its hydroelectric turbines. Sure enough, over there, toward the river-guzzling giant, went the *Ruth D. Jones* and the stern of our fifteen-loaded-barge tow—completely off course. The boat was jumping up and down, I was scared to death, and the big red sign on top of the monolithic concrete building was much closer than I ever wanted to see it.

I sat behind him, frozen to the liar's bench, wondering what he was going to do. Had it been me, I would have had a pool of sweat at my feet and would have long since kicked the pilot's chair across the room. I would have wasted my time cursing and berating myself for getting into that kind of position. But he didn't squander his energy with such useless nonsense. Why he had gotten over there was now no concern of his. The only thing that mattered was the future, of righting the course. In the midst of the panic that the lock men, the deck crew, the chief engineer, and I felt, he rose from the pilot's chair and folded up a piece of paper towel. He turned to me and said, "Son, put this under that coffee pot, would you, please? I'm afraid the dern thing's going to rattle on to the floor."

There you have it! Good housekeeping always prevails. Of course, within a few minutes, the *Ruth D. Jones* slowly started to lift her stern off the power plant, and into the lock we went. Adrian Hargrove didn't tell you, "Keep your cool." He showed you how to do it.

I am not the only person on the river who loved him. There were many people who knew his voice on the radio. His articulation was always smooth, precise, gentle, and trustworthy. In the first few minutes of his watch, some other pilot would invariably call and say, "Adrian, go over to Channel 65." It would be the first of many such calls.

They would chat for a while, the other pilot calling Adrian by name. Finally, Adrian would say, "By golly, and who in the world am I speaking with? Is this Captain Bob?"

"No, Adrian, this is Tim."

"Why, Captain Tim! By gosh, I thought that was your voice. And how in the world have you been getting along? How's your wife?"

"I'm not married, Adrian."

"Oh, that's right. I was thinking that."

And so it would go. No sooner had he cleared with Tim, then Joe would call. Then Larry. Then Steve. He talked to all of them the same, all as good friends. I'm surprised he ever left the pilothouse. But his watches went by quickly. I always loved relieving him in the middle of one of these conversations. That meant he would stay in the chair, and I could relax a few more minutes. They just didn't want to let Adrian go.

I knew of a very few people Adrian was not able to charm. Some were even envious of his knowledge, the respect others had for him, and his bright outlook on life. Not that he never got angry. He often got perturbed with me when we worked together. Mostly, it happened when he felt that I hadn't done my best and had let myself down. He, like any good teacher, was always there to challenge me. Unfortunately, I sometimes didn't push myself as hard as I should have. At those times, he was there to give me that nudge out onto the tightrope.

He was a tough act to follow. Coming on watch after him was like pitching the second game of a doubleheader after Bob Gibson had pitched the first. It was impossible for me to measure up to him. I think that I disappointed him when I left towboating in 1983, though he never told me so. But he was there to help me find a job when I went back to piloting in 1990.

Adrian had a code he lived by, though he never wrote it down or even voiced it. But I saw it in him. When you talk to people, always be polite. You need to be a gentleman. You must respect people, especially when they don't have as much as you do. You need to be there to help and encourage others to do their best. You need to respect nature, especially the river, just the way he did. Perhaps, I will someday live up to it.

When I attended the funeral service, I had no inkling that Adrian's widow, Shirley, would honor me by asking me to be a pallbearer. For an instant, I forgot that I was being treated for a bad back, though I did slip my brace on under my suit. And he would have loved that, too.

As I helped carry his beloved coffin, the United States flag draped ceremoniously across it, I could visualize him under there, smiling.

"Make that shyster carry me across the roughest ground you can find!" he was saying. Then I could hear him talking directly to me. "I hope every darn disc in your whole lumbar pops out of place!" His voice was raspy, thinly disguising its warmth, his friendship, and his sense of humor. That's the way he was. It's the way I'll remember him.

Sleep softly, sweet prince. But oh, don't get too comfortable. I'm going to work real hard to make sure I someday wind up in the same place you are. And buddy, I still know how to kill enough time at Lock 13 to bring you on watch looking at the Clinton Bridge.

### Author's note

I wrote this right after Adrian Hargrove's death in 2000.

# CHAPTER 22

# Government Fleet Blues

I had a job on the levee
I had me a good lookin' woman, she lived in Hughes
You know that pretty little thing
She kept me with the government fleet blues
—Son House, "Levee Camp Moan"

It was an inevitable date with destiny. September of 2009 found me on my way to Memphis to become assistant master of the *M/V Mississippi*.

Satch (may he RIP) was the mate I had the pleasure of working with when I came aboard. "Cool Captain," he informed me, "we got us three little old missions we do 'round these parts. We got to shuttle the commissioners to the meetin's, and we got to tow mat in the revetment season. The third mission is somethin' you'll prepare and train for but won't never have to do—an' that's blow up that levee at Birds Point."

"'Course, Captain, ev'ry few years, when the river starts risin', you'll get them kaboom barges ready to tow up there 'cause you got to get under the Booth Point Bridge 'fore the water gets too

high. Then you'll sit there for a week or two, 'til the water falls
back out, an' come back to Memphis with all the fireworks still
in them barges. The water'll never come up high enough, so's we
gotta blow that levee out."

The levee that he was referring to at Birds Point, Missouri,
is one of the strategic points designed along the Mississippi to
prevent a repeat of 1927. The 1927 flood was the most destructive
and deadly flood ever on the Mississippi, and the River Commission and the corps decided they could never allow it to happen
again. Birds Point is the uppermost in a series of structures that
can be activated to relieve the burden on the levee system during
a flood. It is a levee that can be blown out to allow the river to
expand laterally before being confined by a "setback levee." The
other components of this comprehensive water diversion plan are
the Old River Control Structure, the Morganza Spillway, and the
Bonnet Carre Spillway, all in Louisiana.

My first revetment season, in 2009, went well enough, though we
had to abort it a month early due to unseasonably high water—a
precursor of what was to come. The power of the river has, for
centuries, been capable of crumbling riverbanks and creating
cutoffs. With the hundreds of dikes, or wing dams, now in the
river, the current is even stronger than previously, hitting what
were mud banks. It's not that the corps is against cutoffs. They
simply want to inform the river where and when the cutoff is to
occur. The corps has, indeed, engineered numerous cutoffs and
redesigned channels—but always on their terms. Since the obstrep-
erous, meandering Mississippi does not appreciate this form of
subjugation to man, man must come up with direct methods of
adherence to the rules for the river to follow. Hence, the banks
are lined with the revetment that stabilizes them.

The first style of bank stabilization revetment, instituted around
the turn of the twentieth century, was made of willow trees and
other natural fibers. It was not sturdy enough to withstand the
river's onslaught, not to mention the fact that cutting down all

the timber made the banks erode even faster and diminish the borrow pit of material. Articulated concrete mat is now employed as revetment on every major bend on the Lower Mississippi River (almost one thousand miles). A unique system, originated in Japan, utilizes gantry cranes to lay the revetment from the water's edge out into the river. This system prevents the river from undercutting the bank. The concrete mat is fabricated at three different mat fields along the river, and it is the *M/V Mississippi*'s mission to help deliver barges loaded with mat to the selected jobsite.

The *Mississippi* is the largest river towboat in the world. Though not the most powerful, it is probably the biggest diesel towboat ever built. I say "diesel towboat" because the *Sprague* ("Big Mama," built in 1901) was a steam-powered towboat that was longer. You can authentically call a boat legendary when its length is estimated at 276–315 feet depending on the source, as was the case with the *Sprague*. Anyway, the *Mississippi* is 241 feet long and fifty-eight feet wide. It is *Mississippi* number five, for historical accuracy's sake, preceded since 1882 by three steamboats and one other diesel-powered towboat, the *Mississippi* number four (now a museum in Vicksburg).

When a person enters the main lobby of the *Mississippi*, then proceeds up sixty-eight stairs to the pilothouse, it is also a procession through time—a history lesson about the Mississippi River. Mission statements, awards, and photos grace all the landings on the path upward. It is the flagship of the Corps of Engineers and is used as such by the Mississippi River Commission (established in 1879) at various times of the year to conduct River Commission hearings. It is certainly the only towboat to have a conference room that can seat over 150 people. The boat also has overnight accommodations for over seventy people. River stakeholders are invited on board at these times for "town halls" with the commissioners, whose membership customarily consists of three high-ranking military personnel, including the commanding general, an admiral from NOAA, and three presidentially appointed civilians (two of whom are engineers). The commission is mandated by Congress to conduct these hearings aboard the *M/V Mississippi*.

I feel honored when given the opportunity to serve my country aboard this vessel. Not to mention the fact that the Mississippi River Commission (MRC) trips were interesting since we got to go to rivers that we didn't transit during revetment season. In my ten years, I went up the Ohio to Cincinnati, the Cumberland to Nashville, up the Tennessee to Chattanooga, up the Upper Mississippi to St. Paul, down the Atchafalaya to Morgan City, across the Intracoastal Waterway to Houma, to Bossier City on the Red, and, of course, there was the unforgettable trip up the Arkansas River. We would stop in different towns for stakeholder meetings, and since I was originally a pilot with no ambassadorial duties, I could partake of my favorite pastime when off watch. I love to walk and think. Sometimes, I even leave the thinking part out and just walk. That's what got me into trouble at Dardanelle.

Captain Pete Ciaramitaro had left for private industry, and the corps had appointed me to take his position. In my first MRC low-water inspection trip as master, we had gone up the Arkansas to Muskogee, Oklahoma, and were on our way back down. The Arkansas winds through a valley that is sparsely populated in most places, almost primordial in its darkness after sunset. Our schedule was to stay at Dardanelle Lock for the evening. The corps maintains a service base there, about 130 miles upriver from Little Rock.

We arrived at Dardanelle at a lazy time of the evening, and with the *Mississippi* safely secured, I decided to go for a stroll. Since the commissioners had their own agenda for the night, I didn't think they would need me for a half hour or so. I quickly changed out of my uniform into a T-shirt and shorts.

Leaving the mate and pilot temporarily in charge, I set out on my walkabout at about 9:30 p.m. I took a head lamp with me, expecting an evening black as coal, perfect for a mile or two of exercise. A few hundred yards up a slight hill from the boat, I was suddenly surprised by a parade of headlights. Car after car was coming down that road for some clandestine gathering that the locals were having there. I had not yet received my artificial hip, so I walked with a noticeable limp. There was no shoulder to walk on, and the headlights were blinding me. Every time a car came toward me, I had to shade my eyes and limp into the ditch on the side of the road to keep from becoming Arkansas roadkill. My pleasant walk had begun to seem like another inspired folly. Finally, after ten or fifteen minutes of discomfort, the car procession ceased. I decided, after a few more minutes, to turn around before they all came back out the road to get their second crack at me.

I had just turned back toward the boat when I felt the presence of a car behind me. I moved to the side, but the car did not go around, only paced slowly astern of me. I was reluctant to look back. Then I noticed the blue light flashing in the shadows. It was time for me to stop.

"Whachu doin' out here by yer lonesome, ol' boy?"

"Takin' a stroll, sir," I replied uneasily.

"You ain't from around here, is ya, son?" Before I could answer, "Had some reports of an ol' drunk out stumblin' aroun'. That wouldn't be you, would it?"

"I haven't had a drink in thirty years, sir."

He displayed skepticism concerning that. Passing a field sobriety test on a dark night in rural Arkansas with a severely arthritic hip and lights in your eyes can probe your skills of concentration. It did mine. But it also allowed him to get close enough to smell that I hadn't been drinking.

Then he said, "Let me see your ID."

Oops! I had not taken my wallet with me, hence no ID. I sensed I was rapidly morphing from public nuisance to public enemy.

"Where you come from, man?" He pressed.

Luckily, we were in sight of Dardanelle Lock and the *Mississippi*. The majestic towboat took on the aura of an alien spaceship, towering grandly above the trees at the lock. It was at least twice the size of any towboat on the Arkansas and lit up like a Christmas tree.

I pointed a shaky hand toward the boat. "I'm the captain of that vessel, sir."

The officer looked down his nose at my Iron Maiden T-shirt and camouflage cargo shorts. He shook his head in reflection, deftly searching for a suitable postulate to my claim.

"Do tell? Then I must be Sinbad the Sailor," he finally mused.

I felt my situation sliding inexorably down the hill. What would the general think if the boat couldn't sail in the morning because the captain was in jail? Decades before, in the midst of one of my many blunders while training to become a pilot, Leroy Drury had looked at me and said, "Lee, what was you thinkin'?"

I sheepishly, yet honestly had replied, "Maybe I wasn't." That reflection seemed strangely appropriate at the time.

"Man, why don't we jist follow ya down ta that boat an' see if anybody on there recognizes ya?"

What could I do? But it was a ray of hope and sanity.

The mate was at the landing swapping stories with a couple of deckhands when we arrived back. He smiled furtively to see his captain followed by a squad car.

"This man belong on here?" my suspicious escort queried.

I looked sternly at my crew. The expression on my face advised them that they better not be playin'.

"Yep, that's our captain."

"Give these fine fellas a tour of the boat," I told the mate. Might as well make some friends when you can.

Of course, at breakfast the next morning, the general came down and put his hands on my shoulders.

"We thought we'd have to get a collection to bail you out last night." I hoped he was smiling when he said that.

My crew obviously didn't believe that some stories are better left untold.

But I digress. What I really want to tell you about is what happened in 2011, three years previously.

When the *Mississippi* was on station at Vicksburg Engineer Yard (VEY) in 2010, I walked on a road through the swampy lowlands and arrived at a large parking lot. A white structural framework with black markings was buried into the land there. This pole displays the river stage achieved during various floods on the Mississippi. With the record of over fifty-six feet set during the 1927 apocalypse and standing over forty feet above the contemporaneous waterline, I tried to imagine what this area looked like when inundated with water. It was a staggering thought as that marking lurked several feet above my head. Those things, I thought, only occur in history books.

In 1973, still in my first year on the river, the Mississippi Valley endured the greatest flood since 1927, an auspicious debut for me. I did have to go on top of the pilothouse of the *Julie Ann* to advise the captain regarding clearances as we struggled up through the Eads Bridge and to carry on a somewhat lengthy conversation with another traveler as he walked across the bridge from St. Louis to Illinois. I was a young lad at that time and contemplated knocking out a couple of pull-ups on the tarnished steel arches that hung just a few feet above my head. I would have had time to do that. We were in that bridge, fighting against the snarling current, a lot longer than was safe, though I didn't realize the danger at the time. It was just another story to impress my friends with when I finally got home.

For all I knew, heinous floods happened every spring. I had not yet gained an historical perspective on what it was that I was experiencing.

Any reader interested in the Mississippi River and its significance in American culture owes it to themselves to read *Rising Tide* by John Barry. It chronicles, in detail, the cataclysm of 1927. The river, in effect, became an inland sea inundating 27,000 square miles of land. Several hundred, maybe up to one thousand, people died, though official death records were kept inefficiently, especially those of the African American community. The exodus of Black agricultural workers to the North had already begun, but the flood exacerbated it. Like two world wars, Vietnam, civil rights, and the Depression, the 1927 flood was a benchmark of twentieth-century Americana.

In the winter of 2010–11, to the delight of skiers, Vail Mountain in Colorado received over five hundred inches of snow, more than twice the normal amount. While that snow does not go directly into the Mississippi watershed, it is indicative of the snowfall that inundated the West and the North, much of which did find its way to the Mississippi. The Upper Mississippi in St. Paul, Minnesota, crested in March 2011 at well over flood stage. We knew it was coming. It was inevitable. That's when it started to rain for real around us in Memphis.

Many areas in the mid-Mississippi and Ohio Valleys received six to ten times normal rainfall at the end of April and in the first week of May. In the last week of April, at Ensley Engineer Yard in Memphis, "bomb" barges were loaded with liquid sodium perchlorate and aluminum powder. I was informed that, unless mixed, neither was explosive. However, when packed in tubes together and ignited, they would create the bang that could "activate" Birds Point. "Activate" was a euphemism for blowing it up and flooding 130,000 acres of rich farmland in Missouri.

When the *Mississippi* departed Ensley in the last days of April with the bomb barges in tow, we still had hoped that what we were engaged in was merely preparedness. We had to get the barges above the Booth Point Bridge before the water rose too much for the *Mississippi*'s pilothouse to fit under the bridge steel. We were planning to get the barges up to Hickman, Kentucky, and to "stand by."

We would get the two hundred miles up to Hickman, then, as Satch had so eloquently laid out, sit a spell and wait for the water to fall back out before returning to Memphis.

With a certain segment of the population, our mission, should we indeed activate the floodway, was highly unpopular. And our actions had not gone unnoticed by the press. For that reason, the Coast Guard dispatched four patrol boats to escort us up the river on our mission. Each of these small vessels was manned by three or four coastguardsmen. These ladies and gentlemen were quite well armed, and one of our staterooms on the *Mississippi* became the ordinance room.

The first evening out of Ensley Engineer Yard, I came on watch right in downtown Memphis by the Wolf River. Not surprisingly, I could see more ominous looking clouds coming in from the Arkansas side, where all the bad storms the last few days had initially appeared. Pitchforks ignited, vertically and horizontally, from the sky and came toward me and my bomb barges. Soon, Thor's hammer could be heard, exploding immediately after the lightning. Evil was descending from the heavens, but there was no shelter to be found on a flooding river. Luckily, I made it up a couple of miles above the I-40 Bridge before the storm hit. Normally, a downbound vessel has the right of way, but the rules of the road were altered for our mission. A downbounder slowed for me, and we met below Redman Bar before the maelstrom arrived.

Tornado warnings were in the area, and though I can't say it with any certainty, I believe I might have found me one. Rain and wind attacked the *Mississippi* from no particular direction, and visibility ended just on the outside of the pilothouse glass. At any moment, I expected a full-grown Arkansas cow to come blowing through the windows to take residence with me in the pilothouse. The radar screen was a large blob of green, so I disregarded it as useless. I would ride this one out just with my electronic chart, or so I planned. That's when I looked at the chart to see the boat that I had just met coming back toward me.

Either one of us had gotten turned around, hopefully him, or my antenna for the chart had succumbed to the elements. Uncertain of what was transpiring, I found myself horribly alone and drifting toward the land of discomfort.

I called that boat I had met to ask him if he was still going down on the bridges, but he certainly had problems of his own to occupy his mind. He didn't return the call. If it was a tornado, they usually pass quickly, but the minutes seemed like hours as I waited for the visibility to clear up.

Finally, I began to see a bit of water around the barges. Then some distant riverbanks appeared through the torrential rain. At least they were parallel to me, but was I still going up, or was I headed back down on the bridge? Soon, I could see the entire tow. None of the barges had broken out of their places, heading down the river on their own independent path toward bridge piers. Then, hallelujah, I could see I was still pointed up the river. We were all intact and going in the right direction in life!

That's when I thought about those Coasties with all the weaponry out in those little boats. How could all of them have survived a tornado (if it was such) on a flooding river? Fearful that we had lost at least a few of them, I called them on the VHF radio.

I waited for a tense minute. Then I heard a faint voice return my call, followed directly by a second one. They had all made it. I looked at the clock in the pilothouse and noticed that I was only one hour into my watch. Five more hours like that could damn near kill a man.

As we approached the Booth Point Bridge the next day, I volunteered to repeat my performance from 1973 on top of the pilothouse. We cleared it by a little less than two feet, which was about what we had thought. We took note of what was on the Caruthersville, Missouri, gauge and the vertical clearance markings on the bridge. At some point in time, we would have to come back down through it, and we would be making about fifteen miles per hour with the fair tide behind us. I made up my mind that I would certainly not be on top of the pilothouse when that event took place.

We arrived in Hickman the next day and waited for the commissioners to join us. The *Mississippi* was to become the command center for the flood fight. We had a couple of dry days ahead, which rekindled hope that some random deus ex machina would descend from the heavens and save the day. I am sure Major General Michael Walsh, the president of the commission, wanted that more than any of us did. However, on May 1, it began to rain buckets again, and a dismal pall of realization came over us that the unthinkable was about to take place: the Corps of Engineers, the constructor and protector-in-chief of levees, was about to demolish one and flood out a bunch of people's homes and farmland.

Indeed, all the required data for activation were in place, and no more water from the Tennessee and Cumberland Rivers could be kept back to mitigate the crest. Cairo was being evacuated, and crews were battling the Armageddon by constructing berms and sandbag rings to support levees in Illinois and Kentucky.

The mandated path for General Walsh to follow seemed to be lit, but it was nonetheless an extremely grave decision that he had to make. It is true that the floodway had previously been activated in 1937 and that landholders had been, in effect, prepaid for their land by the government. They had known that this day could come; they just assumed that it never would. The governor and two senators from Missouri pleaded with the corps and the Obama administration but to no avail. They filed suit with the Court of Appeals, but the corps and the commission held all the cards. Even an eleventh-hour appeal to the Supreme Court fell on deaf ears.

Until this flood, the prevailing mission of the Mississippi Valley Division had been to build and rebuild levees and surge barriers in New Orleans and the Gulf states subsequent to Hurricane Katrina. The images of people, particularly those from economically vulnerable populations, begging for rescue from their rooftops was still fresh in everyone's mind. Fertile farmland would be sacrificed, though later renewed; of that, there was no doubt. The wildlife (deer, raccoons, beaver, possum, among others) would also experience extreme carnage. But it was of critical importance that no human beings, either residents or corps workers, be in harm's way when it happened. All of that, I am certain, was on the general's mind as he ordered the evacuation of the floodway.

I did my duty on this mission to the best of my abilities, but my job was miniscule and hardly heroic compared to what others did. My friend Captain Cary Lewis was in charge of one of the smaller boats that had to wind its way in through the floodplain back to the levee. Levees on the Lower Mississippi are sometimes a half mile or more back away from the riverbank. All that batture was flooded so they had to follow a dim path, one barge at a time, back to where the levee sat. It was cold, windy, and raining hard as the workers loaded the explosive material into the vertical pipes where it would sit until zero hour. Cary later shared with me that it was tough, ugly work performed in miserable conditions, with the floodwater, at times, lapping only a few inches from their boots. The people actually on the levee mixing the explosives were clearly the shining stars of that day.

On the night of May 2, we were anchored across from Economy Boat Store, a couple of miles from Birds Point and just downstream from Cairo. General Walsh gave orders for us to proceed down river. The crew pulled anchors, and I turned our tow around when the windows in the pilothouse began to shake so hard that I thought they would burst into shards. I assumed, at the time, that we were being hit by another violent storm, and Pete ran up to the pilothouse concerned I had run over a submerged dike.

It was neither. I turned to see fireballs and flaming embers silhouetted against the night sky. Fire and debris from the explosion billowed hundreds of feet into the air. It was time to get the hell away from there. The rest of the plan was to demolish levees in two more places to allow more water in, then provide an escape path for it back into the river down around New Madrid. In reflection,

2011 was the bicentennial year for the great New Madrid earthquake of 1811, and we were certainly making the earth quake anew. The first Mississippi River steamboat, the *New Orleans*, had made its maiden voyage during that earlier catastrophe. If I thought of that at the time, it was mere coincidence. I was too taken in by the gravity of the moment.

Will history recognize activation of the floodway as an unmitigated success? Certainly, some of the farmers in the area would dispute that. But the engineering, in the most frightening and stressful of times, proved to be effectively designed. In a matter of a couple of hours, river stages began to fall. The figures bandied about are so staggering they are difficult to comprehend. There was 1.8 million cubic feet per second of water coming by the confluence of the Ohio and Mississippi Rivers when the floodway was activated, headed to north of 2 million cfs. The opening drew 400,000 cfs of that water and, almost certainly, saved many mainline levees in the system. There is little doubt that it also saved the city of Cairo, Illinois, from devastation.

The demolition of the outflow levee down at New Madrid the next day was equally memorable and, done during the daylight hours, much more visible. It was time for us to return to Memphis. We did clear the Booth Point Bridge by a few inches making fifteen miles per hour. As the flood moved, like a pig in a python (metaphor courtesy of Mississippi Governor Haley Barbour), down the river, so did the *M/V Mississippi*. We were later involved with the same cast of folks down in Louisiana for the activation of the Morganza Floodway, capping the 2011 flood fight.

In closing, a tiny yet remarkable event occurred after our eventual return to Ensley Engineer Yard that demonstrates the accuracy of military planning. I received a call from the Coast Guard base in Memphis about the ordinance that had been stored in that state room for those weapons taken aboard the small boats. Upon inventory, one round of the hundreds they had in their arsenal on the *Mississippi* had turned up missing. Even today, I wonder if it was lost in one of the small boats during the storm, slipped underneath one of the lockers in the state room, or was buried with some of the other vivid memories of 2011.

# CHAPTER 23

# The Kingfish and the Combat Zone

A river as big as the Mississippi, which is somewhere between 2,300 miles and 2,551 miles in length, depending upon who's measuring, is far too long to be taken as just one river. On its journey from Lake Itasca, in north-central Minnesota, to the Gulf of Mexico, the Mississippi River takes on several characters.

Even a casual student of the river will see at least four rivers: the Wilderness River, from Lake Itasca to St. Anthony Falls; the Upper Navigable River, from St. Anthony Falls to the mouth of the Missouri River; the Middle Mississippi River, from the mouth of the Missouri to Cairo; and the Lower Mississippi River, from Cairo to the Gulf.

For those who have journeyed the Mississippi by water, however, there is a fifth river. It is what professional navigators refer to, not affectionately, as the "Combat Zone." On southbound trips, they begin looking forward to entering the Combat Zone several hours before they arrive. They look forward to this event like eastern Europe looked forward to the invasion of Genghis Khan.

The gateway and symbol of the Zone is the bridge that guards it like Homer's Cerberus guarding the Underworld. It is the Huey P. Long Highway/Railroad Bridge, or the US 190 Bridge—often referred to simply as the Upper Baton Rouge Bridge. This is not to be confused with another Huey P. Long Bridge that spans the river immediately above New Orleans.

The Baton Rouge Bridge, like the man it is named after, has an unsavory reputation among some people. It turns out, however, that the bridge's reputation, at least, is undeserved. That reputation is tied to the Combat Zone and the common navigator's aversion to it.

Boats enter the Zone just above Baton Rouge at a navigation light fittingly called "Devil's Swamp," where pilots must switch their VHF bridge-to-bridge radios from Channel 13 to Channel 67. This switch can be regarded as a rite of passage into the Underworld. Reemergence is a blessed event for most navigators.

Ironically, piloting should become easier below Baton Rouge because the water is much deeper. Instead of a nine-foot channel, a depth of over forty feet is maintained for deep-draft, ocean-going ships. Therein lies a big part of the problem: there is too much traffic, too many docks and fleets, and too many different types of boats.

That's not to say the Combat Zone is a free-for-all. There is a hierarchy, or pecking order, and neophytes must learn it and learn it quickly. At the top of the order are large ships. At the bottom is everybody else. That pecking order is partly based on size. Ocean-going ships are big and fast. Simple survival logic dictates moving out of their way. There is, however, another reason for deference. Much of the economy of the state of Louisiana depends upon the products in the big ships' holds.

At the head of this economic and political zone stands the 190 Huey P. Long Bridge. The navigator coming down on this bridge usually has three spans to choose from. The Baton Rouge, or left descending, is the widest, at 748 feet, but it also has the swiftest current. The Port Allen, or right descending, is considered to be the channel span, but it is closed off by a sandbar in low water. It is also technical in nature to navigate into during high water, necessitating a flanking maneuver by large tows. The center span is 623 feet wide and is difficult to line up for. Too bad none of these three options are any good.

Next to Vicksburg, Mississippi, the Upper Baton Rouge Bridge is the most clobbered bridge on the Lower Mississippi River. It sits immediately below the sharp turn at Wilkinson Point and has several pesky docks right above it and below it. The bridge was hit by a chemical barge one year during high water, forcing the closing of the river and the evacuation of hundreds of nearby residents from their homes.

Still, most of the unsavory reputation of the bridge has little to do with its navigational oddities. It does have to do with the person it was named after.

Huey P. Long was the governor, senator, and some say dictator of Louisiana from 1928 until his assassination on the steps of the state capitol in 1935. Depending on how one viewed him, he was either a twentieth-century Robin Hood—the kingfish who helped the downtrodden—or else a rude and despicable tyrant. Two things, however, are indisputably true about him: he was the quintessential Louisiana politician, and he brought the state into the twentieth century. In his four years of governorship, he oversaw the construction of 1,583 miles of concrete roads, 718 miles of asphalt roads, and 111 bridges—awesome totals for that period of history.

Among the bridges constructed in the 1930s was the Upper Baton Rouge Bridge. Its vertical clearance is 110 feet above the Low Water Reference Plane or 62 feet lower than the I-10 Bridge, a mere four miles downriver. Although 110 feet would be considered a very high bridge on the Upper Mississippi, it is not nearly high enough for tall, ocean-going ships to fit under, especially when high water cuts thirty-five to forty feet off the clearance. The bridge is a barrier across the river for such ships and thus the end of the Combat Zone.

That, the legend goes, is exactly what Huey Long planned when he had the bridge built. According to the legend, Long wanted none of the international business represented by the ships to escape his state, so he dictated that the bridge be built low enough to keep ships from going up to Natchez, Vicksburg, or Memphis. That is, he created the Combat Zone to keep money in his state.

Such a Machiavellian scheme would doubtlessly have appealed to Long. The trouble with the legend is that it's not true. It appears that, as far as Long was concerned, the bridge was just another bridge, if he paid any attention to it at all.

Clara Isenhour, a riverlorian on the *American Queen*, debunked the legend by pointing out a report done by Karen "Toots" Maloy of the *Mississippi Queen*. Maloy wrote:

> In 1996, I had the good fortune to meet the son of C. A. Myers, head designer for the Upper Baton Rouge Bridge. He gave me his father's phone number and I called. Myers was in his 90s at the time and hard of hearing, but he was quite adamant about the real reason the bridge was built so low. He said the Federal Government did not want to maintain a 40-foot channel above Baton Rouge, so there was no need to build the bridge high enough for ships to pass beneath. Myers also informed me that the bridge was completed in 1940. Huey Long was killed in 1935. But did he have a hand in the planning stages? Not according to Myers.

When asked, historians in the Old State Capitol Museum in downtown Baton Rouge said the same thing.

Too bad. Another good legend blown out of the water. History aside, most river pilots are happy the bridge is so low. It keeps the Combat Zone where it is.

And that is exactly where I am heading. I believe I will make the bridge this morning before my watch ends and bring the captain on in a relatively good place. I can go to sleep with a clear conscience and try to get my four hours, then do it all over again. I must keep following the peep light where it leads me—as long as my body is willing.

# A True Friend of the River

David Lobbig was a shining gift to the world, a sincere human being who cared deeply for planet Earth and all the life on it. Anyone who had the good fortune to come in contact with him could put trust in his words. They carried with them a sense of humility and truth. The spiritual realm is a better place with him there, but we, remaining in this life, suffer at his passing.

I first met David in the late 1980s while I was working with the Big Mountain Support Group, taking supplies to the Four Corners region. He, Steve Jones, and I transported food and clothing from St. Louis to the Dine (Navajo) people there. We slept in traditional hogans, shared quiet meals with elders, and herded sheep in a setting so ethereal it seemed of a different galaxy. Entranced by the cloudless night sky in the vastness of the desert, we then set out on a multiday backpacking trip in the Grand Canyon. Upon our return, we shared some other experiences, then I moved from St. Louis, and didn't hear from him again until just a few years ago.

He had become an influential player in the Coalition for the Environment and was also the environmental-life curator for the Missouri History Museum. His bosses had shrewdly tasked him to spearhead the *Mighty Mississippi* exhibit there. Through mutual friend Byron Clemens, he contacted me, and I was honored to do a reading and some other technical augmentation regarding navigation in the display.

In working with him again, I rediscovered traits I had forgotten about him, or maybe he had developed them over the years. What ignited his soul was a curiosity for the world around him. David was all about nature and the work of other humans in that natural setting. In addition, he wanted to know everything he could find out about towboats and commercial navigation. *Ahem*, one of my favorite topics. But I couldn't wear him out. And I'm not just talking about dates and numbers and names. He wanted to know how barges were put together, what the cook prepared for dinner, where the crew slept, what their feelings were about being away from their families, the dirt and grease under their fingernails.

What truly impressed me, though, was his dedication to displaying the pilothouse of the *Golden Eagle* and its disposition subsequent to the dismantling of the exhibit. He felt honor bound to preserve it as a memory of the erstwhile steamer, but particularly to Ruth Ferris, whom I have also written about in this book.

Hagonee, my friend, I will see you around the bend.

# REFERENCES AND CREDITS

Though this book is written as a memoir, there are a few references I would like to mention.

In "An Insignificant Participant in River History," I referenced *A Treasury of Mississippi River Folklore* (1955) by B. A. Botkin; the *Waterways Journal*, thanks to John Shoulberg and his staff; *Way's Steam Towboat Directory* (1990) by F. Way and J. W. Rutter; *Steamboats on the Western Rivers* (1949) by Louis C. Hunter; *Life on the Mississippi* (1883) by Mark Twain; and *The Sultana Tragedy: America's Greatest Maritime Disaster* (1992) by Jerry O. Potter.

In "Government Fleet Blues," I referenced *Divine Providence* (2012) by Charles Camillo, US Army Corps of Engineers.

In addition, earlier versions of several of these stories first appeared in *Big River Magazine*, based in Winona, Minnesota, including:

- "Hubie and the *Hawk*"
- "Sharing a Passion for the River"
- "Deadly Decks"
- "Maneuvering an Elephant in a Bathtub"
- "The Search for the Lock and Dam that Never Was"
- "Just Havin' a License Don't Make a Man No Pilot"
- "Ain't No Such Thing as a Purty Bridge"
- "The Kingfish and the Combat Zone"

I also acknowledge the use of quotes from John Hoover, Keith Norrington, Dennis Kundert, and the late Jim Swift. Thanks to Dave Calvert from the Memphis Regional Examination Center for his information regarding Coast Guard licensing.

# ABOUT THE AUTHOR

**Leo "Lee" Hendrix** began working on the Mississippi in 1972 as a towboat deckhand. He became a pilot on the river in 1976 and stayed with that until 1983. He took a sabbatical from river life in the mid-1980s, working in Europe for an outdoor-education center. Upon returning to the US, he worked with the Student Leadership Environmental Adventure Program (SLEAP), in the St. Louis Public School system. During his years there, he got married, had a family, worked for the North Carolina Outward Bound School, and earned a master's degree in communications from Webster University.

Lee never left the river that flows within him, though, and sought part-time work as a pilot is St. Louis. He later became a captain of passenger vessels and casinos in the Midwest. In 1994, he went to work on passenger steamers, the *American*, *Mississippi*, and *Delta Queens*. To achieve that goal, he was obliged to draw the river from memory, which included over 2,300 miles of the Mississippi and Ohio Rivers.

During that period, Lee began writing stories for *Big River Magazine* in Winona, Minnesota. Several of those are included in this book. He has also had stories published by the *St. Louis Post Dispatch* and the *Waterways Journal*.

In 2009, Lee went to work for the US Army Corps of Engineers as assistant master on the *M/V Mississippi*, the largest river towboat in the world. He was a participant in the 2011 flood fight and later became master of the *M/V Mississippi* in 2014. He retired from the corps in 2018.

After retirement, Lee did not leave his profession. He has done "trip work" for several years and dabbled in writing for a living. Currently, he works on the *American Queen* as riverlorian. When off the boat, he spends most of his time writing and walking his black Lab, Charlie, with his wife, Dianne, around Lake City, Minnesota.